QUALITATIVE ORGANIC CHEMICAL ANALYSIS

QUALITATIVE ORGANIC CHEMICAL ANALYSIS

W. J. Criddle, B.Sc., Ph.D. (Wales)

G. P. Ellis, B.Sc., Ph.D. (Lond.), F.R.I.C.

University of Wales
Institute of Science and Technology
(Designate)

Springer Science+Business Media, LLC

First published by
Butterworth & Co. (Publishers) Ltd.

©

Springer Science+Business Media New York
1967
Originally published by Butterworth & Co. (Publishers) Ltd. in 1967.
Softcover reprint of the hardcover 1st edition 1967

ISBN 978-1-4899-6190-7   ISBN 978-1-4899-6383-3 (eBook)
DOI 10.1007/978-1-4899-6383-3

Suggested U.D.C. number: 543·061: 547
Library of Congress Catalog Card Number 67-26700

Set in 'Monophoto' Times New Roman at the Universities Press, Belfast.

PREFACE

The identification of organic compounds has long been an important part of the training of chemists. The main purpose of this type of work is to familiarize the student with the reactions of organic compounds and to develop his ability to make logical deductions about the nature of the compounds on the basis of his observations. Such training is not possible in organic chemistry except in this type of work; synthetic and quantitative experiments require largely manipulative skill whereas qualitative analysis tests the intellectual *and* practical abilities of the student. In recent years however, there has been a tendency to devote less time to qualitative organic analysis so that time may be found for some of the modern techniques such as spectroscopy and chromatography. We feel, however, that qualitative organic analysis is sound training for the student provided that he applies his theoretical knowledge to the problem and does not mechanically follow a scheme of analysis which leads to the right conclusion without any mental effort on his part.

The aim of the present book is to enable the student to identify organic compounds mainly by virtue of their chemical properties. An important part of this process is presented (in Chapter 2) as a logical sequence of chemical tests which enable the student to determine the functional group(s) present. It is not possible in a book of this size to enlarge on the theoretical basis of each test but the lecturer in charge should supply the necessary background according to the student's ability. These tests are presented in a tabular form to facilitate their use in the laboratory. Ideally the student should study his practical textbooks *before* entering the laboratory so that he has time to consider the principles on which the tests are based. An alternative is for the lecturer to discuss the basis of the analytical scheme in a tutorial class before the student begins his practical work.

The tables of organic compounds and their derivatives contain rather more compounds than are given in many books of this kind. This reflects the improved availability of organic compounds in recent years and enhances the value of the book to advanced students and graduates. It also means that the student sometimes has to use his initiative to select appropriate tests which distinguish compounds of a similar melting or boiling point. An attempt has been made to include many compounds of moderate price that were not readily available a decade or two ago. For these and many others, the original literature has been searched for suitable derivatives in order to provide a selection of these. We ask the indulgence and co-operation of readers in correcting any errors that may have escaped our attention.

PREFACE

We are grateful to many of our students who have helped to verify parts of the work and to Mr. A. H. Henson for criticism and advice on Chapters 1 and 2. We also thank Mrs. J. Criddle and Mrs. G. Ellis for assistance in the preparation of the manuscript.

W. J. C.
G. P. E.

CONTENTS

CONTENTS

1

PRELIMINARY TESTS

1. ELEMENTAL ANALYSIS

The identification of the elements contained in an organic compound is a first and most important step in organic analysis and is best effected by using Lassaigne's test. In this test the organic compound is decomposed by fusion with sodium. The presence of the elements nitrogen, halogens, sulphur or phosphorus in the original compound is then determined by various tests on the product. If nitrogen is present in the compound, fusion with sodium converts it into sodium cyanide which may be identified by the reaction of the cyanide ion in solution. Halide, sulphide or phosphate ions in the product similarly indicate the presence of halogen, sulphur or phosphorus in the compound being analysed.

Lassaigne's test does not provide information on the presence of carbon, hydrogen or oxygen. The vast majority of organic compounds contain carbon and hydrogen and it is usually possible to identify a compound without testing specifically for oxygen although the ferrox test (described below) gives a positive test for the majority of oxygen-containing compounds.

Lassaigne's test

CAUTION: Metallic sodium must be handled with great care as it reacts violently with water and many other compounds. It must never be allowed to come into contact with the skin and protective goggles should be worn.

For solids. Place a piece of metallic sodium (about 2 mm cube) in an ignition tube and heat gently until it is molten. Add the organic compound (about 200 mg) and continue heating gently until the contents of the tube are solid. Then heat more strongly and maintain at red heat for 5 minutes. While the tube is still red hot, plunge it into a 50 ml beaker containing distilled water (15 ml). Boil for 3–4 minutes and filter the solution. Use the filtrate (A) for the tests given below.

For liquids. Heat a piece of metallic sodium (about 2 mm cube) in an ignition tube until the tube is one-third full of sodium vapour. Introduce the liquid (0·2 ml) dropwise into the tube using a dropping-tube. When all the liquid has been added, heat strongly for 5 minutes; then plunge the red-hot tube

1

into a 50 ml beaker containing distilled water (15 ml). Boil for 3–4 minutes and use the filtrate A for the following tests.

Nitrogen

To filtrate A (1 ml) add solid ferrous sulphate (0·5 g) and shake well. If no precipitate appears, the filtrate A is not basic because insufficient sodium was used in the fusion. In such cases, basify the solution with a little 2 N sodium hydroxide solution until a heavy precipitate of ferrous hydroxide is obtained. Boil this for 2 minutes, cool and acidify (test with litmus paper) with 2 N sulphuric acid. A precipitate or coloration varying from deep blue to green indicates the presence of *Nitrogen* in the organic compound.

Sulphur

To filtrate A (1 ml) add freshly prepared sodium nitroprusside solution. A pink to purple coloration indicates the presence of *Sulphur*.

Halogens

To filtrate A (1 ml) add excess of 2 N nitric acid (test with litmus) and if nitrogen or sulphur has been detected by the above tests, boil the solution for 5 minutes to remove hydrogen cyanide or hydrogen sulphide. Boiling is not necessary if nitrogen and sulphur are absent. To the cool solution add silver nitrate solution. A white or yellow precipitate indicates the presence of one or more of *Chlorine, Bromine* or *Iodine* in the organic compound. If this test is positive, the identity of the halogen will be revealed by the following tests, but before these are attempted it is advisable to carry out a blank test on the reagents alone.

If it is known that only one halogen is present it may be identified as follows: to filtrate A acidified with dilute sulphuric acid add chloroform (1 ml) and then chlorine water or 1% sodium hypochlorite solution (2 drops). Shake well and allow the chloroform layer to separate. A brown coloration in the chloroform layer indicates *Bromine*, a purple coloration *Iodine*, while no change in the colour of the chloroform indicates *Chlorine*.

If more than one halogen may be present, the following series of tests should be carried out:

(a) To the filtrate A (2 ml) add an excess of 2 N nitric acid followed by 5% mercuric chloride solution (1 ml) (POISON). A yellow precipitate which changes to orange or red on standing for a few minutes indicates the presence of *Iodine*. If a high concentration of iodide ions is present in the solution, the precipitate is orange or red immediately.

(b) To filtrate A (2 ml) add an equal volume of dichromate oxidizing mixture and boil gently for 2 minutes. Test the vapours produced with a filter paper dipped in freshly prepared Schiff's reagent. A purple coloration indicates the presence of *Bromine*.

2

(c) To filtrate A add an excess of 2 N nitric acid and then silver nitrate solution. Filter off the precipitate and treat it with an excess of a solution consisting of four volumes of saturated ammonium carbonate solution and one volume of ammonia solution (0·88). If a precipitate remains, filter this off and acidify the filtrate with dilute nitric acid. A white precipitate indicates the presence of *Chlorine*. It should be noted that silver bromide is slightly soluble in the solution used above. A faint precipitate obtained during this test should therefore be ignored.

(d) Acidify a further portion of filtrate A (2 ml) with acetic acid; bring the solution to the boil and cool. Add a drop of this solution to a piece of filter paper dipped in zirconium-alizarin solution (1% ethanolic alizarin and 0·4% aqueous zirconium nitrate) and allow the paper to dry. A red to yellow colour change indicates the presence of *Fluorine*.

Phosphorus

Treat a further portion of filtrate A (2 ml) with concentrated nitric acid (0·5 ml) followed by a 5% solution of ammonium molybdate. Warm but *do not boil*. A yellow precipitate indicates the presence of *Phosphorus*.

Ferrox test for oxygen

Grind together equal weights of potassium thiocyanate and ferric alum. Place the mixture (about 100 mg) in a test tube and add the organic compound directly if a liquid or as a saturated solution in benzene or chloroform if a solid. A purple coloration in the organic layer indicates the presence of oxygen in the compound. This test is specific for oxygen only if nitrogen and sulphur are absent.

2. IGNITION

Place some of the organic compound (0·1 g) on a spatula or in an ignition tube and heat until ignition occurs. Remove from the flame and observe the ignition characteristics. A clear flame indicates an aliphatic compound while a smoky flame is characteristic of aromatic and some unsaturated compounds. Continue the ignition until no further change occurs; the presence of a residue shows that the original compound contains a metal atom and the residue should be examined by the standard inorganic procedures for the identification of metals. In the majority of cases, a flame test carried out on the residue, which should be acidified by addition of concentrated hydrochloric acid (1 drop), will be sufficient to identify the metal atom present. It is sometimes possible to recognize the odour of the vapours released during ignition; some compounds (for example, carbohydrates, aliphatic hydroxy acids and their salts) char readily while others (for example, benzoic acid) sublime unchanged.

3

3. COLOUR AND ODOUR

The majority of organic compounds are colourless when pure but some compounds become discoloured on standing due to the formation of small amounts of coloured impurities. If a pure sample is coloured it is likely to contain one or more chromophoric groups, for example, nitro, nitroso or azo group, or to be a quinone or to have an extended conjugated system of four or more double bonds. The nitro group on its own confers very little if any colour on a compound but if an auxochromic substituent such as a hydroxyl or amino group is also present, the very pale yellow colour is intensified. An indication of the colour of many compounds is given in the tables of melting points at the end of this book.

Some organic compounds have characteristic odours which can be used tentatively to guide us in organic analysis but since smell cannot often be described in words, the student's best approach should be to try to 'memorize' the smell of some common compounds.

4. DETERMINATION OF PHYSICAL CONSTANTS

Melting point

Seal off a standard melting-point tube at one end in a Bunsen flame. Introduce the sample to a depth of about 2 mm at the sealed end of the tube (rubbing the tube with a nail file often facilitates this operation) and attach the tube by means of a rubber band to a short-immersion 360° thermometer with the sample at the same level as the thermometer bulb. Clamp the thermometer at a depth of about 3 cm in an oil-bath (the rubber band should be above the oil level) and then heat gently over a gauze with a micro-burner. Stir continuously while adjusting the burner to give a heating rate not greater than 5 degrees per minute. The temperature at which a meniscus forms from the molten sample is the required melting point.

Notes: (*a*) It is often better to determine the approximate melting point of the sample first and then redetermine the melting point more accurately on a second sample by slowing down the heating rate to 2 degrees per minute as the approximate melting point is approached.

(*b*) For samples which are thermally unstable it is preferable to determine the approximate melting point in the ordinary way and then to determine the accurate value by heating the oil-bath to within 10 degrees of the rough value before inserting the sample; the temperature is then raised at 2 degrees per minute until the compound melts. This reduces the period during which the sample is being heated and thus minimizes thermal decomposition.

(*c*) Where an electrically heated melting point apparatus is available, it should be used according to the instructions provided with the apparatus, although paragraphs (*a*) and (*b*) are valid for any melting point determination.

4. DETERMINATION OF PHYSICAL CONSTANTS

Mixed melting point

When two different compounds are mixed together and the melting point of the mixture is determined, it is found that melting begins at a temperature several degrees below that of the lower-melting pure compound. This technique of *mixed melting points* may therefore be used to determine whether or not the two samples are identical. A depression in the melting point of a sample when mixed with another indicates that the compounds are different. The correct procedure for such a determination is as follows: grind together equal weights of the known and unknown materials and introduce the mixture into a melting-point tube. Place a little of each of the two pure compounds in two other tubes and determine the melting points of all three simultaneously. If the mixture melts more than 5 degrees below the melting point of either of the pure samples, the latter are different. If they are identical, the three samples will melt at the same temperature.

Boiling point

Siwoloboff's method

Take a glass tube about 10 cm long and 0·5 cm internal diameter and a standard melting-point tube and seal each at one end only. Introduce the liquid under examination (0·5 ml) into the larger tube and place in it the melting-point tube, *open end* in the liquid. Attach the tubes to a short-immersion 360° thermometer with the liquid at the same level as the thermometer bulb. Immerse the thermometer in a liquid paraffin bath to a depth of 3 cm. Heat the bath at a constant rate with continuous stirring until a steady stream of bubbles emerges from the lower end of the smaller tube. Stop heating at this point and note the temperature at which the liquid rises rapidly into the smaller tube. This is the boiling point of the liquid. If the sample is impure (for example, contains a small amount of water) Siwoloboff's method will give misleading values. It is then better to remove the impurity by fractional distillation or by thorough drying with a desiccant.

Solubility in various solvents

The solubility of an organic compound in water, ether, 2 N hydrochloric acid and 2 N sodium hydroxide can often furnish useful information about the nature of the compound. However, the presence of more than one functional group may have such a profound effect on the solubility that it is often impossible to make deductions about the functional groups present from solubility data. For example, resorcinol is extremely soluble in water but the introduction of a butyl group into the 4-position gives a compound which is only slightly soluble. Even positional isomers sometimes differ greatly in their solubility, for example, the solubilities of *o*-, *m*- and *p*-dihydroxybenzene in water at 20° are 45, 210 and 7 per cent respectively. The following table of solubilities should

5

therefore be used with caution, and it is likely to be most accurate for mono-functional compounds. It gives an indication of the solubility of various types of organic compounds in ether, water, 2 N hydrochloric acid and 2 N sodium hydroxide. The compounds are grouped together according to the elements which are identified in the Lassaigne test. A + sign means that the compounds of that class have a solubility in the particular solvent in excess of 5 per cent. A lower solubility is indicated by a − sign. A class of compounds whose members vary greatly in solubility in a particular solvent is shown as ±.

Table of solubilities

	Ether	Water	2 N HCl	2 N NaOH	Comments
Compounds containing C, H, O, *Metal*					
Carboxylic acids					
aliphatic	+	+[a]	+[a]	+	[a] Insol. if >4 C atoms.
aromatic	+	−	−	+	
metal salts	−	+	±	+	
Phenols					
monohydric	+	−[b]	−	+	[b] Phenol is sol.
polyhydric	−	+[c]	±	+	[c] Phloroglucinol is insol.
phenoxides	−	+	±	+	
Aldehydes and Ketones					
aliphatic	+	−[d]	−[d]	−[d]	[d] Sol. if <4 C atoms.
aromatic	+	−	−	−	
Acetals	+	±	−[e]	±	[e] Hydrolysis occurs
Alcohols	+	+[f]	+[f]	+[f]	[f] Sol. if <4 C atoms.
Carbohydrates	−	+	+	+	
Polyols	−	+	+	+	
Esters	+	−[g]	−[h]	−[h]	[g] Sol. if <4 C atoms.
					[h] May hydrolyse to sol. products
Anhydrides	+	±[i]	−	+	[i] Aliphatic, +; aromatic, −.
Lactones	+	−[j]	−	+	[j] γ-Butyrolactone is sol.
Quinones	+	−	−	+	
Ethers	+	−	−	−	
Hydrocarbons	+	−	−	−	
Compounds containing C, H, N (O)					
Amines					
pri. aliphatic	+	+	+	−	
s-aliphatic	+	−[k]	+	−[k]	[k] Sol. if <4 C atoms.
t-aliphatic	+	±[l]	+	−	[l] Variable.
pri. aromatic	+	−	+	−	
s-aromatic	+	−	+	−	
t-aromatic	+	−	+	−	
Amides	−	−[m]	−	−	[m] Sol. if <6 C atoms.
N-Substituted amides	+	−	−	−	
Imides	−	−	−	+	

6

Table of solubilities (continued)

	Ether	Water	2 N HCl	2 N NaOH	Comments
Ammonium salts	−.	+	+[n]	+	[n] Depends on solubility of free acid.
Nitrocompounds	+	−	−	−	
Amino-acids	−	±[o]	±[o]	±[o]	[o] Variable.
Nitroso-, azo- and azoxy- compounds	+	−	−	−	
Hydrazides	−	−	+	−	
Hydrazones	+	−	−	−	
Arylhydrazines	+	−	+	−	
Phenylhydrazones	+	−[p]	−	−	[p] Most sugar hydrazones (except that of mannose) are soluble.
Oximes	+	−	−	−	
Semicarbazones	+	−	−	−	

Compounds containing C, H, S (O)

	Ether	Water	2 N HCl	2 N NaOH	Comments
Sulphonic acids	−	+	+.	+	
Thiols and thiophenols	+	−	−	+	

Compounds containing C, H, P (O)

	Ether	Water	2 N HCl	2 N NaOH	Comments
Phosphate esters	+	−	−	−	

Compounds containing C, H, Halogen (O)

	Ether	Water	2 N HCl	2 N NaOH	Comments
Alkyl and aryl halides	+	−	−	−	
Acyl halides	+	+[q]	+[q]	+[q]	[q] Decomposed, alkyl compounds rapidly.

Compounds containing C, H, Halogen, N (O)

	Ether	Water	2 N HCl	2 N NaOH	Comments
Quaternary ammonium salts	−	+	+	+	
Hydrohalides of organic bases	−	+	+	+	

Compounds containing C, H, Halogen, S (O)

	Ether	Water	2 N HCl	2 N NaOH	Comments
Sulphonyl halides	+	−	−	−	

Compounds containing C, H, N, S (O)

	Ether	Water	2 N HCl	2 N NaOH	Comments
Thioamides	−[r]	−[r]	−	+	[r] Sol. if <3 C atoms.
Sulphates of organic bases	−	+	+	−	
Sulphonamides	+	−	−	+	

FUNCTIONAL GROUP ANALYSIS

When the elements which are present in an organic compound have been determined, it is then necessary to ascertain how these are arranged in the molecule, that is, what *functional group(s)* it contains. For this purpose, use is made of the chemical reactions which are characteristic of each function. The saturated hydrocarbons contain no functional group but the number of such groups in other compounds may be one (monofunctional) or more (difunctional, trifunctional, etc.).

The tests described below are grouped according to the element(s) detected in the Lassaigne test. Compounds containing only C, H (and possibly O) should be examined by the tests given in *Table I*. If *one* of the elements halogen, N, P or S is present, the tests in *Tables II–V* should take precedence. Where *two or more* of these elements are present, the initial tests should be for the composite groups containing the relevant elements (*Tables VI–IX*); for example, if N and S have been detected, the presence of a sulphonamide group ($-SO_2 \cdot NH_2$) among others should be investigated as described in *Table VIII*. If these tests are negative, a search should be made for the presence of groups containing the individual elements, that is, sulphur-containing functions (*Table IV*) and nitrogen-containing functions (*Table III*). The scheme of analysis outlined in this chapter is capable of being used to identify compounds of varying degrees of difficulty. Students beginning qualitative analysis should only test for those groups shown in heavy print. The remaining groups are included for more advanced students.

When a functional group has been identified, it should be possible to correlate this with the results of the preliminary tests so that examination of the appropriate melting point table (p. 46) will narrow the choice to one or more possible compounds. In such cases, consideration of the structure of the likely compounds may suggest further functional group tests to differentiate between them and thus enable the compound to be identified. For example, an organic compound, m.p. 122°, containing C, H, N, (O) was shown to be a primary aromatic amine. Reference to *Table 9* shows that it may be one of three compounds having a melting point close to 122°, namely, *m*-aminophenol, 2,4-dimethyl-5-nitroaniline or *p*-aminobenzophenone. These may be distinguished by the appropriate tests for the second functional group, that is, phenolic, nitro and keto group respectively. The identification may then be confirmed by the preparation of one or more of the derivatives given in *Table 9*.

A series of tests contained within heavy horizontal lines is for related functions. In many cases, if the first test in a series is negative, the subsequent tests in that particular series may be omitted.

Table I. Compounds containing C, H, (O)

Test	Observation	Inference
1. Prepare a saturated solution in 50% aqueous ethanol and add an equal volume of 5% sodium bicarbonate solution. Shake well but do not heat.	Carbon dioxide evolved.	**Carboxylic acid**
2. Dissolve the organic compound (0·5 g) in water or aqueous ethanol (5 ml) and add N sodium hydroxide solution (1 drop). Test the solution with B.D.H. '1014' or similar indicator.	Green colour.	**Carboxylic acid or phenol**
3a. Dissolve in *water* and add neutral aqueous ferric chloride.	Wide range of colours. (See *Table 30*)	**Phenol** or enol (α-hydroxy acids give yellow colorations)
3b. Dissolve in *methanol* and add methanolic ferric chloride (anhydrous ferric chloride (5 g) dissolved in methanol (100 ml)).		
3c. If test 3a or 3b is positive, prepare a cold solution of mercurous nitrate in 2 N nitric acid and add the organic compound.	No precipitate.	**Phenol**
	Immediate formation of a grey precipitate of mercury.	Enol
4. Add a solution of 2,4-dinitrophenylhydrazine in *either* 5 N hydrochloric acid *or* ethanolic sulphuric or phosphoric acid.	Red to yellow precipitate formed.	**Aldehyde, ketone** or acetal
4a. If test 4 above is positive, add aqueous ammonia to a 5% silver nitrate solution until the precipitate just dissolves and add the organic compound. A little ethanol may be added if necessary to dissolve the compound. Warm gently but *do not boil*.	Silver mirror formed.	**Aldehyde**
	No reaction.	**Ketone** or acetal
4b. If test 4a indicates the presence of a ketone, mix the organic compound with a little 10% iodine-potassium iodide solution and add 2 N sodium hydroxide solution until the iodine colour is discharged. Warm gently, add a further drop of sodium hydroxide solution and sufficient iodine solution to give a permanent yellow colour.	Yellow precipitate of iodoform.	**Ketone containing the $CH_3 \cdot CO$— group**

2*

9

Table I (continued)

Test	Observation	Inference
4c. If test 4 is positive but test 4a is negative, warm the compound with N hydrochloric acid for ·15 minutes. Just neutralize with aqueous ammonia and then apply test 4a above.	Silver mirror formed.	Acetal
5a. Add chromium trioxide in 5 N sulphuric acid and warm gently.	Colour change from red to yellow.	**Alcohol**
5b. Dissolve the compound in water or dioxan and add a few drops of 4% ceric nitrate solution in 2 N nitric acid.	Red coloration.	
6. If tests 5a and 5b are positive, dissolve the compound in water and add 10% ethanolic 1-naphthol followed by careful addition of concentrated sulphuric acid (Molisch's test).	Violet coloration at liquid interface.	**Carbohydrate**
6a. If test 6 above is positive, mix equal volumes of 0·1 M copper sulphate and 0·1 M sodium potassium tartrate. Add 0·1 M sodium hydroxide until the precipitate formed just dissolves (Fehling's solution). Add this solution to an aqueous solution of the organic compound. Heat on a boiling-water bath for 5 minutes.	Red precipitate of cuprous oxide formed.	**Reducing sugar**
6b. Dissolve the carbohydrate in water and add a 6% solution of cupric acetate in 1% aqueous acetic acid (Barfoed's reagent). Boil gently for *up to 2 minutes*.	Red precipitate of cuprous oxide formed.	**Monosaccharide**
6c. Add a 4% solution of benzidine in glacial acetic acid (Tauber's reagent) to the carbohydrate. Boil for 1 minute.	Yellow to brown coloration.	**Hexose**
	Cherry-red coloration.	Pentose
7a. Mix equal volumes of saturated methanolic hydroxylamine hydrochloride and saturated methanolic potassium hydroxide. Add the organic compound and heat on a boiling-water bath. Ethanol may be added if necessary to solublize the mixture. Cool, acidify with 0·2 N hydrochloric acid, and add a few drops of 5% aqueous ferric chloride.	Red to violet colour obtained.	**Ester,** lactone or **anhydride**

Table I (continued)

Test	Observation	Inference
7b. Dissolve in ethanol and add N ethanolic potassium hydroxide (1 drop) and phenolphthalein (1 drop). Prepare a similar mixture but omit the compound under examination. Heat on a boiling-water bath.	Pink colour fades in test solution.	As for test 7a.
7c. If test 7a or 7b is positive, dissolve in benzene or chloroform and add aniline. Warm gently for 1–2 minutes.	Precipitate formed.	**Anhydride**
8a. Visual examination.	Compound has a red to yellow colour when pure.	**Quinone**
8b. Treat with 2 N sodium hydroxide.	Pronounced intensification of the original colour.	
9a. Warm with concentrated sulphuric acid.	Compound dissolves completely without charring.	**Ether**
9b. If tests 1–8 are negative, the 'ferrox test' (p. 3) should be applied.	Violet coloration.	
10a. Add 1% potassium permanganate solution and shake well.	Purple colour is discharged rapidly in the cold.	**Alkene** or alkyne
10b. Dissolve in carbon tetrachloride and add two or three drops of a 5% solution of bromine in carbon tetrachloride.	Red-brown bromine colour is discharged without the evolution of hydrogen bromide.	
11. All the tests listed above are negative.		**Other hydrocarbons**

Table II. **Compounds containing C, H, Halogen, (O)**

Test	Observation	Inference
1a. To the compound add aqueous silver nitrate acidified with 2 N nitric acid.	White to yellow precipitate.	**Acyl halide**

Table II (continued)

Test	Observation	Inference
1b. If test 1a is negative, shake the halogen compound with 2% ethanolic silver nitrate solution.	White to yellow precipitate formed in the cold or on slight warming.	**Alkyl halide**
1c. If test 1b is negative, reflux the compound with 10% ethanolic potassium hydroxide solution for 10 minutes. Cool, acidify with 2 N nitric acid and add silver nitrate.	No precipitate.	**Aryl** or vinylic **halide** or chloroform

Table III. **Compounds containing C, H, N, (O)**

Test	Observation	Inference
1a. Dissolve* in 2 N hydrochloric acid at room temperature, cool to 5° in ice and add 5% aqueous sodium nitrite.	Effervescence; nitrogen is evolved and a clear solution is obtained.	**Primary aliphatic amine, amino-acid, amide** or semicarbazide
* A few weakly basic amines require concentrated hydrochloric acid; in extreme cases, the amine should be dissolved in the minimum of ethanol and a little concentrated sulphuric acid should be added and the solution cooled in ice.	No effervescence; clear solution obtained.	**Primary aromatic amine, tertiary amine** or hydrazide
	No effervescence; cloudy solution (or emulsion) formed.	**Secondary amine** or arylhydrazine
1b. If a clear solution is obtained in test 1a above, treat the solution with 2-naphthol (5% solution in 2 N sodium hydroxide).	No coloration; ignore white to yellow precipitates.	**Primary aliphatic amine, tertiary amine** (see † below), **amino-acid, amide, semicarbazone, hydrazide** or **semicarbazide**
	Bright red to dark brown precipitate.	**Primary aromatic amine** or arylhydrazine
	† Dark brown solution obtained.	† **Tertiary aromatic amine** with vacant p-position

12

Table III (continued)

Test	Observation	Inference
1c. Dissolve in 2 N hydrochloric acid at room temperature and add Fehling's solution (*Table I*, test 6a) in excess, i.e. until solution is alkaline. Heat on a boiling-water bath for 5 minutes.	Red precipitate of cuprous oxide.	Substituted hydrazine, hydrazide or semicarbazide.
1d. If test 1c above is negative, warm with concentrated hydrochloric acid for 1 minute, and then treat as in test 1c above.	Red precipitate of cuprous oxide.	Hydrazone, osazone, oxime or semicarbazide
2. Add 8 N aqueous sodium hydroxide. Heat strongly.	Ammonia evolved readily.	**Ammonium salt, amide** or **imide**
	Ammonia evolved slowly.	**Nitrile**
2a. Mix with sulphur in a dry test-tube and heat gently. Test the vapours produced with an absorbent strip treated with 1% aqueous ferric nitrate.	Red stain formed on paper.	**Nitrile** or oxime
2b. If test 2 is positive, add cold aqueous 2 N sodium hydroxide.	Compound dissolves and ammonia is evolved.	**Ammonium salt**
	Compound dissolves but no ammonia evolved.	**Imide, lower amide or nitrile**
	Compound insoluble, no ammonia evolved.	**Higher amide or nitrile**
2c. If an imide is suspected, mix saturated solutions of the organic compound and potassium hydroxide in methanol.	White precipitate formed.	**Imide**
3. Mix with 70% sulphuric acid and reflux for 10 minutes. Cool in ice to 5° and add aqueous sodium nitrite (5%) followed by alkaline 2-naphthol (see tests 1a and 1b above).	Bright red to dark brown precipitate.	*N*-**Substituted amide**
4a. To an aqueous solution of ferrous sulphate add a few drops of 2 N sodium hydroxide followed by the organic compound. Prepare a similar mixture but omit the organic compound. Heat on a boiling-water bath.	Grey to green precipitate in test solution changes to brown.	**Nitro-compound,** but may be nitroso-, azo-, or azoxy-compound

13

Table III (*continued*)

Test	Observation	Inference
4b. Dissolve the compound in acetone and add 5% titanous chloride solution; warm gently.	Mauve colour is discharged within 2 minutes.	As for test 4a.
5a. If test 4a or 4b is positive, reduce the organic compound as follows: add tin and 7 N hydrochloric acid and warm, with continual shaking, for 15 minutes. Filter the mixture, cool to 5° in ice and add aqueous sodium nitrite (5%) and then alkaline 2-naphthol (see tests 1a and 1b above).	Red to dark brown precipitate.	**Aromatic nitro-compound**
	No coloured precipitate; ignore white to yellow precipitate.	**Aliphatic nitro-compound**
5b. Add aqueous 2 N sodium hydroxide.	Intense yellow to orange coloration.	**Nitrophenol**
5c. Warm with a little phenol and concentrated sulphuric acid and then add water.	Blue to green coloration changing to red on addition of water.	Nitroso-compound
5d. If test 5c above is positive, add a 5% solution of potassium iodide in 2 N sulphuric acid to the organic compound.	Immediate liberation of iodine.	C-Nitroso-compound
	Slow liberation of iodine.	N-Nitroso-compound
5e. Reduce with tin and hydrochloric as described in test 5a above.	Various amines formed. Examine as in tests 1a and 1b above.	Azo-, azoxy- or NN'-di-substituted hydrazine
6. Dissolve in water and add 1% methanolic ninhydrin. Warm gently.	Violet coloration.	**α- or β-Amino-acid**

Table IV. Compounds containing C, H, S, (O)

Test	Observation	Inference
1. Add water, shake well and test with litmus paper.	Readily soluble with acid reaction.	**Sulphonic acid**
2. Odour.	Unpleasant and penetrating.	Thiol or thiophenol (also impure thioether)

Table IV (*continued*)

Test	Observation	Inference
2a. If test 2 above is positive, dissolve in ethanol and add solid sodium nitrite followed by 2 N sulphuric acid.	Red coloration.	Primary or secondary thiol
	Green coloration changing to red on standing.	Thiophenol
	No coloration.	Thioether

Table V. **Compounds containing C, H, P, (O)**

Test	Observation	Inference
1. Reflux with 30% aqueous sodium hydroxide for 20 minutes and then distil off all the volatile material. Acidify the residue with 2 N sulphuric acid, extract with ether and add concentrated nitric acid and ammonium molybdate to the aqueous phase. Warm but do not boil.	Yellow precipitate obtained.	Phosphate ester

Table VI. **Compounds containing C, H, Halogen, N, (O)**

Test	Observation	Inference
1. Dissolve in water; add excess nitric acid and then aqueous silver nitrate.	White to yellow precipitate formed.	**Hydrohalide of a base** or quaternary ammonium halide
1a. If test 1 above is positive, add excess alkali and extract the mixture with ether. Dry the extract over anhydrous sodium sulphate and evaporate the ether on a water bath.	Residue obtained; examine as described in *Table III*, tests 1a, 1b and 1c.	**Hydrohalide of a base**
	No residue.	Quaternary ammonium halide

Table VII. Compounds containing C, H, Halogen, S, (O)

Test	Observation	Inference
Heat with water for 10 minutes and cool. Then (a) Acidify with 2 N nitric acid and add aqueous silver nitrate.	White to yellow precipitate.	**Sulphonyl halide**
(b) Add tin and hydrochloric acid.	Unpleasant odour of thiol or thiophenol.	

Table VIII. Compounds containing C, H, N, S, (O)

Test	Observation	Inference
Heat with solid potassium hydroxide.	Ammonia evolved.	**Sulphonamide**
	Amine evolved. Examine as described in *Table III*, tests 1a and 1b.	*N*-substituted sulphonamide
Heat with 2 N sulphuric acid.	Hydrogen sulphide evolved.	Thioamide

Table IX. Compounds containing C, H, N, P, (O)

Test	Observation	Inference
Dissolve in water and treat with concentrated nitric acid and aqueous ammonium molybdate. Warm, but do not boil.	Yellow precipitate obtained.	Phosphate of a base

THE SEPARATION OF ORGANIC MIXTURES

A mixture of organic compounds may be in the solid or liquid form or may consist of a solid dissolved or suspended in a liquid. If a solid and a liquid are present it is usually unwise to expect separation to be accomplished by filtration because the liquid phase almost certainly contains some dissolved solid and traces of the liquid component may be difficult to remove from the solid compound. The methods of isolating pure samples of the components from a mixture may be either physical or chemical. The physical method consists of fractional distillation and is applicable only if there is a wide difference between the boiling points of the two compounds and provided that an azeotrope is not formed. The chemical method of separating two compounds depends on their differing solubility in water, ether, dilute acid or alkali.

The procedures described below should be followed in the order given and should be successful for the great majority of mixtures.

1. If the mixture is a liquid, it should be placed in a small flask equipped with a stillhead, thermometer and condenser. Heat the flask carefully; observe whether a liquid distils, and if it does, note the temperature and continue the distillation until the temperature at the stillhead falls, indicating that all the liquid boiling at that temperature has distilled over. The lower-boiling component of the mixture will now be in the receiver flask.

2. If the mixture is a liquid which cannot be separated according to paragraph 1, or if the mixture is a solid, test its solubility in ether. Most organic compounds are soluble but amongst those which have a low solubility are the following: carbohydrates, amino-acids, sulphonic acids, salts of amines, metal salts of carboxylic and sulphonic acids, some aromatic polybasic acids, some amides and ureas, and polyhydroxy compounds.

Liquid mixtures: proceed to test (b) below.

Solid mixtures: if the mixture dissolves completely, proceed to test (b) below. If an insoluble part remains, continue as described in (a). If both components are insoluble, use the sequence of tests given in paragraph 5 below.

(a) Filter off the undissolved solid, keep the ethereal filtrate (A). Allow the solid collected to dry off in air or over slight heat.

Take the filtrate A and if all the mixture did not dissolve in ether initially, evaporate the ether over a hot water bath. If a residue (solid or liquid) is obtained, a separation has been achieved by virtue of the ether-solubility of one of the two compounds.

(b) If all the mixture dissolved in ether, place the solution in a tap funnel. Extract this with 10% sodium hydroxide solution, separate the basic extract from the ether layer (B) into a small flask and acidify with 10% hydrochloric acid solution. The appearance of a solid or an oil or of cloudiness may indicate the presence of a carboxylic acid or a phenolic compound. To this mixture add an excess of solid sodium bicarbonate in small amounts with constant stirring until the solution is no longer acid to litmus. Extract the solution with ether; the ethereal layer will contain a phenolic compound if present while the aqueous layer will contain a carboxylic or sulphonic acid if present.* Acidify this aqueous solution with dilute hydrochloric acid. A solid carboxylic acid may separate out and should be filtered off and dried. If a liquid acid is present, some solid calcium chloride should be added with thorough shaking until the water is almost saturated with it. The acid is then extracted with ether and isolated in the usual way.

Take the ether layer B and extract it with 10% hydrochloric acid solution. Separate the acid layer (keep the ethereal solution (C)) and basify it with dilute sodium hydroxide. If an oil or a solid separates, extract this twice with ether. Dry the extract with sodium sulphate, allow to stand for 10 minutes in a stoppered flask, evaporate the ether over a hot-water bath and the residue if any will be the basic component.

The ethereal layer C contains a neutral compound if present in the original mixture; it should be dried with sodium sulphate, filtered and the solvent distilled. The most common classes of neutral compounds are: hydrocarbons, ethers, halides, alcohols, aldehydes, ketones, amides, nitriles, esters, unreactive anhydrides and nitro compounds.

3. The procedure given in paragraph 2 above will not separate two neutral compounds. If nothing is extracted from the ethereal solution by dilute acid and dilute alkali, the presence of two neutral compounds should be suspected. If one of these is a carbonyl compound which forms a bisulphite adduct, this may be removed as its addition product. Prepare a 40% solution of sodium metabisulphite in water and add one fifth of its volume of ethanol. If some of the salt separates, filter the solution and to the filtrate (12 ml) add the mixture (4 g). Shake thoroughly and allow to cool. The crystalline adduct if formed should be filtered off (keep the filtrate (D)) and decomposed by careful distillation with an excess of sodium carbonate solution. The carbonyl compound if volatile is found in the distillate; otherwise it will have to be extracted out of the mixture in the distilling flask with ether and isolated in the usual way.

* Some nitro- and halogeno-phenols are sufficiently acidic to release carbon dioxide from sodium bicarbonate. Therefore an aromatic compound which contains nitrogen or a halogen and which behaves as carboxylic acid in this separation should be tested for phenolic properties.

The other neutral component will be found in the filtrate D and should be isolated by removing any ethanol that may remain and then extracted with ether.

4. A few compounds which are very soluble in water and in ether can be troublesome to isolate from a mixture. For example, resorcinol mixed with an amine or a carbonyl compound would not be readily separated by the procedure given in paragraph 2. Resorcinol is about equally soluble in water as in ether and although it would be extracted out of ether with several portions of dilute sodium hydroxide solution, acidification of this extract will not result in separation of the compound because of its solubility in water. A mixture consisting of an amine and resorcinol may be separated as follows:

Dissolve the mixture in an excess of dilute hydrochloric acid; test the solution with litmus paper. Place the solution in a tap funnel and extract it with three portions of ether each equal to about one half the volume of the mixture. Retain the acidic solution (E). Dry the ethereal extract with anhydrous sodium sulphate, filter and distil off the solvent. The residue should be tested for a phenolic group. Basify the acidic solution E with sodium hydroxide solution, extract with ether. Dry the extract and distil off the solvent. A residue indicates the presence of an amine.

A water-soluble phenol or polyhydric phenol and a carbonyl compound are best separated by converting the latter into its bisulphite compound (see paragraph 3) or its semicarbazone and regenerating it by heating with twice its weight of oxalic acid and ten times its weight of water in a distilling flask. The carbonyl compound, if low boiling, may be distilled and collected, or if high boiling, may be extracted with ether from the residue in the distilling flask and characterized in the usual way.

5. The procedures given above will not be effective if both components are insoluble in ether. In such cases the following suggestions should be useful:

(a) Test the mixture to see if one component is soluble in water. If this is so, filter off the insoluble compound and evaporate the filtrate to dryness on a small Bunsen flame. Overheating may cause decomposition or charring. If a carbohydrate is present a syrup will often be obtained which may be difficult to crystallize but tests should be made on the syrup to determine its nature.

(b) Should both components be insoluble in water or both soluble in water, repeat procedure 5(a) using methanol.

(c) If the above tests have failed to effect a separation, the presence of any two of the following should be suspected: polyhydric alcohol, carbohydrate, metallic salt of a carboxylic acid or a salt of an organic base. Dissolve the mixture in 10% hydrochloric acid and if a solid precipitates, collect on a filter, wash with water and dry carefully. This may be a free acid, probably aromatic. Formation of an oil will indicate a higher aliphatic acid. The aqueous layer may contain a soluble polyol or carbohydrate.

19

If no solid or oil is obtained, extract the solution several times with ether and evaporate the extract on a water bath. A residue may be a lower aliphatic acid. If nothing was extracted by the ether, make the solution basic with 10% sodium hydroxide solution; a salt of an organic base if present will lead to the formation of the free base which should be extracted with ether and isolated in the usual way. The aqueous layer may contain a polyol or a carbohydrate together with sodium chloride or sulphate. Removal of the inorganic ions by passing the solution through a mixed-bed ion-exchange column should give an aqueous solution of the polyol or carbohydrate.

4

PREPARATION OF DERIVATIVES

Before an organic compound can be identified with certainty, it is necessary that it be converted to at least one other sharp-melting solid compound called a 'derivative'; liquid derivatives are not considered suitable. A derivative which melts above 80° should be chosen wherever possible because of the reluctance of some low-melting compounds to crystallize especially if traces of solvent are present. The compound is identified by cross reference of its melting or boiling point with the melting point of one of its derivatives.

Purification of derivatives

When preparing a derivative, it is necessary that the product be obtained in a pure form, thus ensuring an accurate melting point. The procedure for recrystallization is as follows: the crude material is dissolved in the minimum of hot solvent, filtered if necessary *without* suction, and the solution obtained allowed to cool slowly. The crystalline product is collected in a Buchner funnel, washed with a small volume of ice-cold solvent, dried thoroughly and the melting point determined. This procedure should be repeated until no further increase in melting point is observed.

If the crude material is badly discoloured, it is often advantageous to add a little activated charcoal to the solution which should then be boiled gently for up to 5 minutes. The hot solution is filtered without suction and the filtrate allowed to cool slowly.

The choice of a suitable solvent for recrystallization is of considerable importance in the purification of derivatives. Ideally, the compound should have a high solubility in the hot solvent and a low solubility in the cold solvent. If two or more solvents meet this requirement, that with the lowest boiling point should be chosen to facilitate removal from the solid product. It is sometimes impossible to find a satisfactory solvent and then the technique of mixed solvents is often of value. The derivative is dissolved in a slight excess of hot solvent and a second solvent which is miscible with the first solvent but in which the derivative is insoluble is carefully added with shaking and warming until a faint cloudiness persists. This should disappear on boiling and the solution is then allowed to cool slowly.

Experimental details are given below for preparing the derivatives listed in Chapter 5 for the various classes of organic compounds.

21

ACETALS

(a) Hydrolysis to aldehyde and alcohol

$$RCH(OAlk)_2 \xrightarrow{\text{aq. acid}} RCHO + 2AlkOH$$

Treat the acetal (0·5 g) with N hydrochloric acid (5 ml) and reflux for up to 30 minutes. If the resulting solution is homogeneous, divide into two equal portions and characterize the aldehyde as the 2,4-dinitrophenylhydrazone (see *Aldehydes*, p. 23) and the alcohol as the potassium alkyl xanthate as follows: To a portion from the hydrolysis add solid potassium hydroxide (7 g) and cool to 40°. Transfer to a separating funnel and add carbon disulphide (3 ml) (INFLAMMABLE) and acetone (3 ml). Mix cautiously and then shake vigorously for 15 minutes. Allow to settle, discard the lower layer and filter the remaining solution through glass wool. Precipitate the xanthate with ether and recrystallize from ethanol.

If two layers separate, they should be treated individually as described above.

ALCOHOLS

(a) Acetate

$$AlkOH + (CH_3 \cdot CO)_2O \rightarrow AlkO \cdot CO \cdot CH_3 + CH_3 \cdot CO_2H$$

To the alcohol (0·5 g) add anhydrous sodium acetate (0·5 g) and acetic anhydride (3 ml). Reflux for 20 minutes and pour into water (25 ml). Stir until a solid is obtained, filter this off and wash well with water. Recrystallize from ethanol.

(b) Benzoate and toluene-*p*-sulphonate (Schotten-Baumann method)

$$AlkOH + C_6H_5 \cdot COCl \rightarrow AlkO \cdot CO \cdot C_6H_5 + HCl$$

$$AlkOH + p\text{-}CH_3 \cdot C_6H_4 \cdot SO_2Cl \rightarrow AlkO \cdot SO_2 \cdot C_6H_4 \cdot p\text{-}CH_3 + HCl$$

Dissolve the alcohol (0·5 g) in 2 N sodium hydroxide (10 ml) and add benzoyl chloride (1 ml) or finely powdered toluene-*p*-sulphonyl chloride (1 g). Shake vigorously in a stoppered tube until a solid is obtained. Filter this off, wash well with water and recrystallize from ethanol.

(c) *p*-Nitrobenzoate and 3,5-dinitrobenzoate

$$AlkOH + ArCOCl \rightarrow AlkO \cdot COAr + HCl$$

where $Ar = p\text{-}NO_2 \cdot C_6H_4—$ or $3,5\text{-}(NO_2)_2 \cdot C_6H_3—$.

22

Dissolve the alcohol (0·5 g) in dry pyridine (5 ml) and add *p*-nitrobenzoyl chloride (1 g) or 3,5-dinitrobenzoyl chloride (1·3 g). Reflux for 30 minutes and pour into 2 N hydrochloric acid (40 ml). Separate the solid (sometimes an oil is formed) and stir with N sodium carbonate solution (10 ml). Filter off the solid obtained and recrystallize from ethanol, aqueous ethanol or benzene.

(d) **Hydrogen 3-nitrophthalate**

Heat a mixture of the alcohol (0·5 g) and 3-nitrophthalic anhydride (0·5 g) until a liquid is obtained. Continue heating for a further 15 minutes, cool, and recrystallize the solid obtained from water or aqueous ethanol. If the original alcohol has a boiling point in excess of 150°, it is advisable to dissolve the mixture in toluene (2–5 ml) and reflux for up to 30 minutes. If no solid product is obtained on cooling, precipitate the product by addition of light petroleum (60–80°).

(e) **Oxidation to carboxylic acid**

$$RCH_2 \cdot OH \xrightarrow{[O]} RCO_2H$$

Treat the alcohol (1 g) with a chromic acid oxidation mixture (10 ml) consisting of 50% sodium dichromate solution in 12 N sulphuric acid. Reflux until the colour has changed from red to green and add more oxidizing mixture. Continue this procedure until no further discharge of the red colour is observed. Cool the solution, filter off the resulting product and wash well with 2 N sulphuric acid and then with water. Dissolve the product in sodium carbonate solution, filter, and acidify with 2 N sulphuric acid. Filter off the solid, wash well with water and recrystallize from water or ethanol.

ALDEHYDES

(a) **2,4-Dinitrophenylhydrazone**

$$RCHO + 2,4\text{-}(NO_2)_2C_6H_3 \cdot NH \cdot NH_2 \rightarrow RCH{:}N \cdot NH \cdot C_6H_3(NO_2)_2 + H_2O$$

To the compound (0·5 g) dissolved in ethanol (0·5 ml) add a solution (2 ml) of 2,4-dinitrophenylhydrazine (see below) and boil for 2 minutes. Filter off

23

the resulting precipitate, wash with a little cold ethanol and recrystallize from ethanol, acetic acid, ethyl acetate or chloroform.

Preparation of reagent

(i) Prepare a saturated solution in either 5 N hydrochloric acid or 5 N sulphuric acid. (ii) Dissolve the 2,4-dinitrophenylhydrazine (2 g) in methanol (30 ml) and water (10 ml). Add concentrated sulphuric acid (4 ml) cautiously with shaking. Cool and filter if necessary. (iii) Dissolve 2,4-dinitrophenyl-hydrazine in a solution of 85% phosphoric acid (60 ml) in ethanol (40 ml). Heat gently if necessary.

(b) Semicarbazone

$$RCHO + H_2N \cdot NH \cdot CO \cdot NH_2 \rightarrow RCH{:}N \cdot NH \cdot CO \cdot NH_2 + H_2O$$

To a solution of the compound (0·5 g) in water (2 ml) add hydrated sodium acetate (0·75 g) and semicarbazide hydrochloride (0·5 g). Add ethanol drop-wise if the solution is not completely clear but care should be taken to add the minimum amount of ethanol otherwise sodium chloride may be precipitated. Heat for up to 10 minutes on a boiling-water bath, cool and filter. Recrystallize the product from ethanol, water, benzene or glacial acetic acid.

(c) Oxime

$$RCHO + H_2N \cdot OH \rightarrow RCH{:}N \cdot OH + H_2O$$

Dissolve hydroxylamine hydrochloride (0·5 g) in water (3 ml) and add hydrated sodium acetate (0·5 g) followed by the aldehyde (0·5 g). Heat on a boiling-water bath and add ethanol dropwise, if necessary, to clear the solution. Continue heating for 1–2 hours and allow to cool. Filter off the solid and recrystallize from ethanol.

(d) *p*-Nitrophenylhydrazone (and phenylhydrazone)

$$RCHO + p\text{-}NO_2 \cdot C_6H_4 \cdot NH \cdot NH_2 \rightarrow RCH{:}N \cdot NH \cdot C_6H_4 \cdot p\text{-}NO_2 + H_2O$$

Prepare a solution of *p*-nitrophenylhydrazine (0·5 g) in ethanol (15 ml) and glacial acetic acid (0·5 ml) and add the organic compound (0·5 g) to the solution. Reflux for 10 minutes, cool and recrystallize the solid product from ethanol. Occasionally, no solid is obtained on cooling. In such cases, the solution should be reheated and water added until a faint cloudiness is observed. Recrystallize the product using this technique (mixed solvents).

For the preparation of phenylhydrazones, replace the *p*-nitrophenylhydrazine with phenylhydrazine (0·5 g).

24

(e) Dimethone

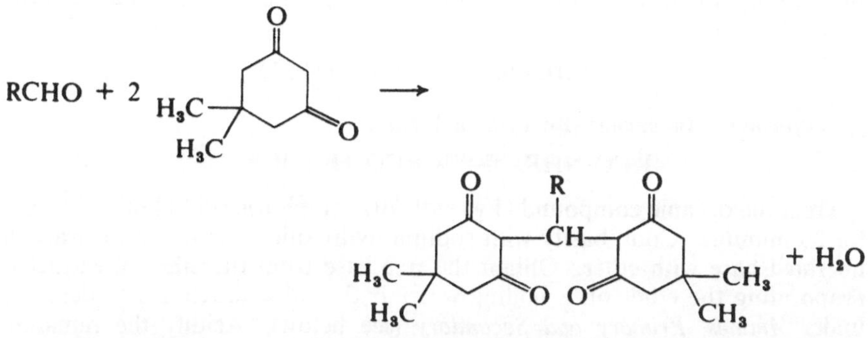

RCHO + 2 →

+ H_2O

Treat the aldehyde (0·5 g) with a saturated aqueous solution of dimedone (20 ml). A little ethanol may be added to dissolve the compound if necessary. If no precipitate forms within 2 minutes, warm the solution for 5 minutes, cool in ice and filter off the product. Recrystallize from aqueous ethanol or ethanol.

AMIDES, IMIDES, UREAS AND GUANIDINES

(a) Xanthyl derivative

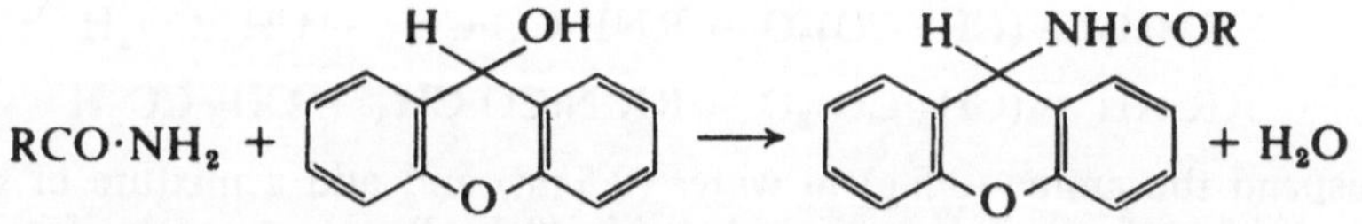

RCO·NH₂ + → + H_2O

To the organic compound (0·5 g) add a 7% solution of xanthydrol in glacial acetic acid (7 ml) and reflux for up to 30 minutes. Add water (5 ml) and allow to cool. Filter off the solid product and recrystallize from aqueous dioxan or acetic acid.

(b) Hydrolysis to carboxylic acid

$$RCO·NH_2 + NaOH \rightarrow RCO_2Na + NH_3 \xrightarrow{aq.\ acid} RCO_2H$$

Reflux the organic compound (0·5 g) with an excess of 6 N sodium hydroxide solution until no further evolution of ammonia is detectable. Acidify the

25

3

resulting solution with concentrated hydrochloric acid and filter off the solid product obtained. Wash well with water and recrystallize from water, aqueous ethanol or ethanol.

AMIDES, N-SUBSTITUTED

(a) Hydrolysis to carboxylic acid and amine

$$RCO \cdot NHR' \xrightarrow{\text{aq. acid}} RCO_2H + R'NH_2$$

Treat the organic compound (1 g) with 70% sulphuric acid (4 ml) and reflux for 30 minutes. Cool, basify with sodium hydroxide solution and extract the liberated base with ether. Obtain the free base from the ethereal extract by evaporating the ether on a boiling-water bath and characterize as described under *Amines, Primary and Secondary* (see below). Acidify the remaining aqueous solution to Congo Red with hydrochloric acid and filter off any solid produced. Recrystallize from water, aqueous ethanol or ethanol to obtain the pure carboxylic acid and characterize as described under *Carboxylic Acids* (p. 31). If no solid separates, saturate the solution with sodium chloride and extract with ether. Evaporate the ether and characterize the residue as described under *Carboxylic Acids* (p. 31).

AMINES, PRIMARY AND SECONDARY

(a) Acetyl derivative

$$RNH_2 + (CH_3 \cdot CO)_2O \rightarrow RNH \cdot CO \cdot CH_3 + CH_3 \cdot CO_2H$$

$$RR'NH + (CH_3 \cdot CO)_2O \rightarrow RR'N \cdot CO \cdot CH_3 + CH_3 \cdot CO_2H$$

Suspend the amine (0·5 g) in water (0·5 ml) and add a mixture of glacial acetic acid (0·5 ml) and acetic anhydride (0·5 ml). Heat gently if reaction does not occur immediately. Cool and filter off any solid which separates. If no solid is obtained, neutralize the solution with saturated sodium carbonate solution. Filter off the solid obtained. In either case, recrystallize the product from water or aqueous ethanol.

(b) Benzoyl, benzenesulphonyl and toluene-p-sulphonyl derivatives

$$RNH_2 + C_6H_5 \cdot COCl \rightarrow RNH \cdot CO \cdot C_6H_5 + HCl$$

$$RR'NH + C_6H_5 \cdot COCl \rightarrow RR'N \cdot CO \cdot C_6H_5 + HCl$$

$$RNH_2 + ArSO_2Cl \rightarrow RNH \cdot SO_2Ar + HCl$$

$$RR'NH + ArSO_2Cl \rightarrow RR'N \cdot SO_2Ar + HCl$$

where $Ar = C_6H_5$— or $p\text{-}CH_3C_6H_4$—.

Suspend the amine (0·5 g) in 2 N sodium hydroxide solution (10 ml) and add benzoyl chloride (1 ml) *or* benzenesulphonyl chloride (1 g) *or* toluene-*p*-sulphonyl chloride (1 g). Shake vigorously in a stoppered tube until a solid separates. Filter this off, wash well with water and recrystallize from ethanol.

(c) 2,4-Dinitrophenyl derivative

$$RNH_2 + 2,4\text{-}(NO_2)_2C_6H_3Cl \rightarrow 2,4\text{-}(NO_2)_2C_6H_3\cdot NHR + HCl$$

$$RR'NH + 2,4\text{-}(NO_2)_2C_6H_3Cl \rightarrow 2,4\text{-}(NO_2)_2C_6H_3\cdot NRR' + HCl$$

Treat the amine (0·5 g) with an equimolar proportion of 2,4-dinitrochlorobenzene (CAUTION: skin irritant) and anhydrous sodium acetate (1 g); heat on a boiling-water bath for up to 30 minutes. Cool, and add cold ethanol (3–4 ml). Filter off the solid obtained and recrystallize from ethanol.

(d) Picrate

$$RR'R''N + 2,4,6\text{-}(NO_2)_3C_6H_2\cdot OH \rightarrow [RR'R''NH]^+[2,4,6\text{-}(NO_2)_3C_6H_2O]^-$$

where R' and/or R'' may be a hydrogen atom.

Dissolve the amine (0·5 g) in ethanol (2 ml) and treat with a saturated ethanolic solution of picric acid (3 ml). Warm gently for 1 minute and allow to cool. Recrystallize the product from ethanol.

AMINES, TERTIARY

(a) Methiodide

$$RR'R''N + CH_3I \rightarrow [RR'R''N\cdot CH_3]^+[I]^-$$

Add methyl iodide (0·5 ml) to the dry amine (0·5 g) at room temperature and allow to stand for 5 minutes. Reflux on a boiling-water bath for a further 5 minutes and then cool in ice. Filter off the solid product (scratch with a glass rod if no solid is obtained) and recrystallize from ethanol or acetone.

(b) Picrate

$$RR'R''N + 2,4,6\text{-}(NO_2)_3C_6H_2\cdot OH \rightarrow [RR'R''NH]^+[2,4,6\text{-}(NO_2)_3C_6H_2O]^-$$

Prepare as described under *Amines, Primary and Secondary*.

(c) *p*-Nitroso derivative (for dialkyl tertiary amines with vacant *p*-position in aryl group).

$$(Alk)_2N\text{—}\langle\ \rangle + HONO \longrightarrow (Alk)_2N\text{—}\langle\ \rangle\text{—}NO + H_2O$$

Dissolve the amine (0·5 ml) in 2 N hydrochloric acid (4 ml) and cool in ice to 5°. Add dropwise 20% sodium nitrite solution (2 ml) and allow to stand in the cold for 5 minutes. Basify with 2 N sodium hydroxide and extract with chloroform. Dry the extract over anhydrous sodium sulphate and precipitate the derivative by addition of carbon tetrachloride. Filter off the product and recrystallize from ether (INFLAMMABLE).

(d) Methyl toluene-*p*-sulphonate salt

$$RR'R''N + p\text{-}CH_3 \cdot C_6H_4 \cdot SO_2 \cdot O \cdot CH_3 \rightarrow$$
$$[RR'R''N \cdot CH_3]^+ [p\text{-}CH_3 \cdot C_6H_4 \cdot SO_2 \cdot O]^-$$

To the amine (0·2 ml) add methyl toluene-*p*-sulphonate (0·3 g) and benzene (1 ml) or isopropyl ether (1 ml). Reflux on a water bath for 20 minutes and cool. Decant the ether (crystals should remain) and add methanol (1 ml) and ethyl acetate (5 ml) for recrystallization.

AMINO-ACIDS

(a) Benzoyl, 3,5-dinitrobenzoyl and toluene-*p*-sulphonyl derivatives

$$\begin{array}{l} CO_2H \\ | \\ R \cdot NH_2 \end{array} + ArCOCl \rightarrow \begin{array}{l} CO_2H \\ | \\ R \cdot NH \cdot COAr \end{array} + HCl$$

where $Ar = C_6H_5-$ or $3,5\text{-}(NO_2)_2C_6H_3-$.

$$\begin{array}{l} CO_2H \\ | \\ R \cdot NH_2 \end{array} + p\text{-}CH_3 \cdot C_6H_4 \cdot SO_2Cl \rightarrow \begin{array}{l} CO_2H \\ | \\ R \cdot NH \cdot SO_2 \cdot C_6H_4 \cdot p\text{-}CH_3 \end{array}$$

Prepare as described under *Amines, Primary and Secondary* (using 3,5-dinitrobenzoyl chloride (1 g) to prepare the 3,5-dinitrobenzoyl derivative). In each case acidify the solution with 2 N hydrochloric acid when the reaction is complete. Filter off the solid obtained and recrystallize from water, aqueous ethanol or ethanol.

(b) Picrate

$$\begin{array}{l} CO_2H \\ | \\ R \cdot NH_2 \end{array} + 2,4,6\text{-}(NO_2)_3C_6H_2 \cdot OH \rightarrow \left[\begin{array}{l} CO_2H \\ | \\ R \cdot NH_3 \end{array} \right]^+ [2,4,6\text{-}(NO_2)_3C_6H_2O]^-$$

Prepare as described under *Amines, Primary and Secondary* (p. 26).

28

(c) Acetyl derivative

$$\underset{\underset{\textstyle R\cdot NH_2}{|}}{CO_2H} + (CH_3\cdot CO)_2O \rightarrow \underset{\underset{\textstyle R\cdot NH\cdot CO\cdot CH_3}{|}}{CO_2H} + CH_3\cdot CO_2H$$

Prepare as described under *Amines, Primary and Secondary* (p. 26).

CARBOHYDRATES

(a) β-Acetate, e.g. of D-glucose:

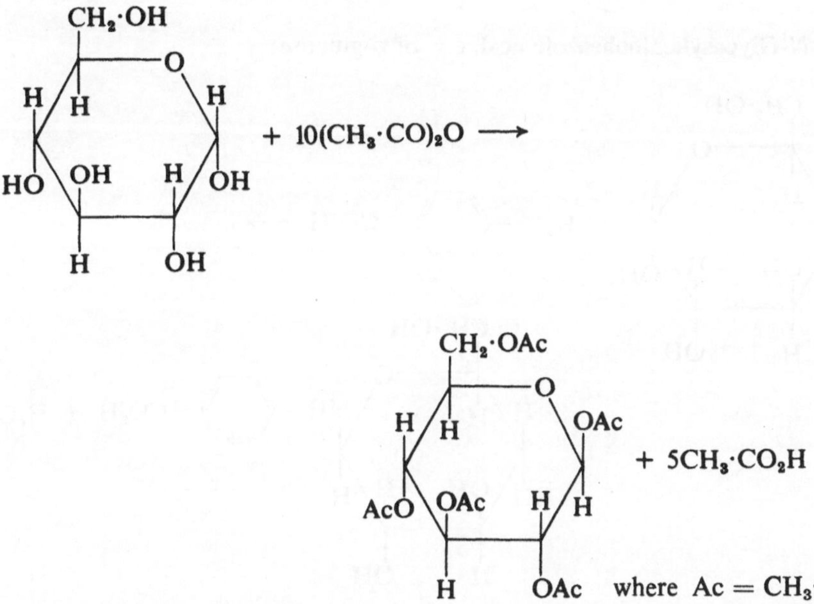

$$+ \ 10(CH_3\cdot CO)_2O \longrightarrow$$

$$+ \ 5CH_3\cdot CO_2H$$

where $Ac = CH_3\cdot CO-$

Treat the carbohydrate (0·5 g) with anhydrous sodium acetate (0·5 g) and acetic anhydride (3 ml). Heat on a boiling-water bath for 90 minutes and pour the product into water (25 ml). Filter off the solid obtained after stirring, wash well with water and recrystallize from ethanol. If an oil is obtained, decant the water and induce crystallization by scratching with a glass rod.

29

(b) **Benzoate** (of glucose and fructose only), e.g. of D-glucose:

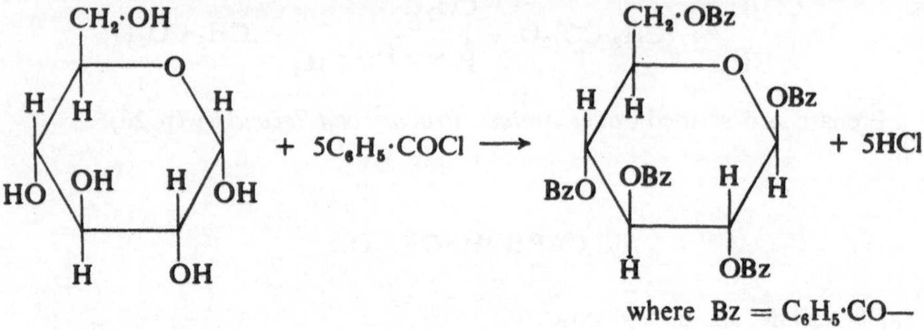

$+ 5C_6H_5 \cdot COCl \longrightarrow$

$+ 5HCl$

where $Bz = C_6H_5 \cdot CO-$

Prepare as described under *Alcohols* (p. 22).

(c) **p-N-Glycosylaminobenzoic acid,** e.g. of D-glucose:

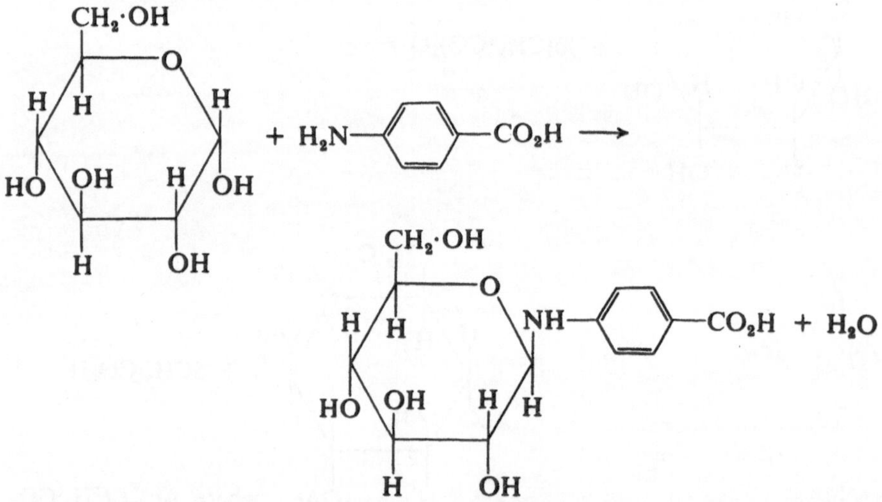

$+ H_2N-\!\!\!\bigcirc\!\!\!-CO_2H \longrightarrow$

$-NH-\!\!\!\bigcirc\!\!\!-CO_2H + H_2O$

Heat the carbohydrate (1 g) with water (not more than 0·5 ml) on a boiling-water bath. When most of the sugar has dissolved, add p-aminobenzoic acid (1 g) in three portions. Continue heating for not more than 4 minutes. Remove the reaction mixture from the water bath and add methanol (4 ml). Cool in ice if necessary and filter off the solid. Wash it with a small volume of cold methanol and dry at room temperature in air or under vacuum. The product may be recrystallized from ethanol if desired. The melting point of this derivative is best determined by introducing the sample into the apparatus which has been pre-heated to 120°, or to 170° if an approximate determination of melting point indicated that the compound melted above 180°.

(d) **Osazone**

$$
\begin{array}{c}
\text{CHO} \\
\mid \\
\text{CH·OH} \\
\mid \\
\text{R}
\end{array}
+ 2C_6H_5 \cdot NH \cdot NH_2 \rightarrow
\begin{array}{c}
\text{CH:N·NH·C}_6\text{H}_5 \\
\mid \\
\text{C:N·NH·C}_6\text{H}_5 \\
\mid \\
\text{R}
\end{array}
+ 2H_2O
$$

Dissolve the carbohydrate (1 g) in water (5 ml) and add phenylhydrazine (1 ml) and glacial acetic acid (1 ml) in water (3 ml). Heat the mixture on a boiling-water bath for 30 minutes and allow to cool. Filter off the product, wash well with cold water and recrystallize from ethanol.

CARBOXYLIC ACIDS

(a) **Amide, anilide and p-toluidide**

$$
RCO_2H \xrightarrow{SOCl_2} RCOCl
\begin{cases}
\xrightarrow{NH_3} RCO \cdot NH_2 \\
\xrightarrow{C_6H_5 \cdot NH_2} RCO \cdot NH \cdot C_6H_5 \\
\xrightarrow{p\text{-}CH_3 \cdot C_6H_4 \cdot NH_2} RCO \cdot NH \cdot C_6H_4 \cdot p\text{-}CH_3
\end{cases}
$$

To the acid (1 g) add thionyl chloride (2 ml) and heat on a boiling-water bath until no further reaction occurs (about 30 minutes). Distil off the excess of thionyl chloride and treat the residue with ammonia, aniline or p-toluidine (1 ml). Filter off the solid product and recrystallize from water, aqueous ethanol or ethanol.

(b) *p-Bromophenacyl ester and p-phenylphenacyl ester*

$$RCO_2H + ArCO \cdot CH_2Br \rightarrow RCO \cdot O \cdot CH_2 \cdot COAr + HBr$$

where Ar = $p\text{-}BrC_6H_4$— or $p\text{-}C_6H_5 \cdot C_6H_4$—.

Prepare a solution of the acid (1 g) in an equivalent amount of sodium hydroxide, make *slightly* acid to litmus by adding a few drops of 2 N hydrochloric acid and add p-bromophenacyl bromide (1 g) *or* p-phenylphenacyl bromide (1 g) (CAUTION: these compounds are eye and skin irritants) in ethanolic solution. Heat to boiling, adding more ethanol if solution is not complete. Continue refluxing for 1, 2 or 3 hours depending on whether the acid is mono-, di- or tri- basic. Cool and filter the product. Recrystallize from ethanol, aqueous ethanol or benzene.

ENOLS

Semicarbazone and 2,4-dinitrophenylhydrazone

$$RC\!:\!CHR' \rightleftharpoons RC\!\cdot\!CH_2R' \xrightarrow{\ H_2N\cdot NH\cdot CO\cdot NH_2\ } RC\!\cdot\!CH_2R' + H_2O$$
$$\underset{OH}{} \quad \underset{O}{} \quad \underset{N\cdot NH\cdot CO\cdot NH_2}{}$$

$$\downarrow{\scriptstyle 2,4\text{-}(NO_2)_2\cdot C_6H_3\cdot NH\cdot NH_2}$$

$$RC\!\cdot\!CH_2R'$$
$$2,4\text{-}(NO_2)_2C_6H_3\cdot NH\!\cdot\!N$$

Prepare as described under *Aldehydes* (p. 23).

ESTERS

(a) Hydrolysis

$$RCO_2R' + H_2O \xrightarrow{\ HO^-\ } RCO_2H + R'OH$$

Various methods are available for the hydrolysis of esters to the parent acid and alcohol or phenol depending on the ease of hydrolysis of the ester and the boiling point of R′OH. Three methods are described below which make use of potassium hydroxide in different solvents, viz., water, ethanol and diethylene glycol.

Aqueous alkali—Reflux the ester (5 g) with 30% aqueous potassium hydroxide (40 ml) until hydrolysis is complete, normally indicated by a change in appearance or odour of the reaction mixture. If a solid separates (usually a sparingly soluble salt of the acid component), filter, wash well with water and characterize as the *p*-bromo- or *p*-phenylphenacyl ester as described under *Carboxylic Acids* (p. 31). If a liquid separates, this may be an insoluble alcohol. In such cases, extract with ether, dry the ether extract over anhydrous sodium sulphate and evaporate the ether on a water bath. Characterize the compound obtained as described under *Alcohols* (p. 22).

Where a homogeneous solution is obtained, saturate with solid potassium carbonate and extract with ether. Dry the ether extract over anhydrous sodium sulphate and evaporate the ether. Characterize the residue as described under *Alcohols* (p. 22). Acidify the aqueous solution with hydrochloric acid and filter off any solid product. Wash well with water and characterize as described under either *Carboxylic Acids* (p. 31) or *Phenols* (p. 41). See Note 1. If no solid was obtained, saturate with calcium chloride and extract with ether. Evaporate the ether and characterize the residue as described under *Carboxylic Acids* (p. 31) or *Phenols* (p. 41). See Note 1.

Methanolic alkali—(for esters of high-boiling alcohols and phenols). Reflux the ester (5 g) with 20% methanolic potassium hydroxide (40 ml) until hydrolysis is complete. If a solid separates, filter off, wash well with methanol and characterize as described under *Carboxylic Acids* (p. 31) or *Phenols* (p. 41). See Note 1. Carefully distil the excess methanol from the filtrate and characterize the residue as described under *Alcohols* (p. 22) or *Phenols* (p. 41). If a homogeneous solution is obtained after hydrolysis, distil off the bulk of the methanol on a water bath, cool and extract the residue with ether. Dry the ether extract over anhydrous sodium sulphate, evaporate the ether and characterize the residue as described under *Alcohols* (p. 22). Characterize the residue remaining as described under *Carboxylic Acids* (p. 31).

Alkali in diethylene glycol—(for esters resistant to hydrolysis). To the ester (5 g) add a solution consisting of potassium hydroxide (2 g) in diethylene glycol (10 ml) and water (2 ml). Reflux the mixture for 5 minutes. Distil off all the volatile material (water and alcohol) and saturate this distillate with potassium carbonate before extracting with ether. Dry the ether extract, evaporate under *Alcohols* (p. 22). If no alcohol is detected, the residue remaining after ether extraction will contain a carboxylic acid and phenol. Dissolve in water and acidify with hydrochloric acid. Filter off any solid obtained and characterize as described under *Carboxylic Acids* (p. 31) or *Phenols* (p. 41). See Note 1. The filtrate will contain the remaining component which should be characterized in the same manner.

Note 1. Where carboxylic acids and phenols of similar solubility in the solvents used are produced together, they may be separated as follows: acidify the residue (if a potassium salt) and saturate the solution with calcium chloride. Extract with ether and treat the ether extract with 5% sodium bicarbonate solution. Separate the ether layer and dry over anhydrous sodium sulphate. Evaporate the ether and characterize the residue as described under *Phenols* (p. 41). Acidify the bicarbonate extract, saturate with calcium chloride and extract with ether. Dry the ether extract over anhydrous sodium sulphate and evaporate the ether. Characterize the residue as described under *Carboxylic Acids* (p. 31).

ETHERS

(a) **Nitro derivative**

where $n = 1, 2$ or 3

Nitration is sometimes a hazardous procedure and must always be undertaken with caution. Various nitration methods are available depending on the ease with which the particular compound may be nitrated. These methods are described below and the preferred method is indicated in the melting point tables.

(i) Mix equal volumes of concentrated sulphuric acid and concentrated nitric acid (5 ml) and add the organic compound (0·5 g). Maintain the temperature at 25° by cooling and shaking until the reaction is complete. If no reaction occurs, the mixture should be heated cautiously to start the reaction. Pour the resulting material into water (50 ml) and stir. Filter off the solid which is formed.

(ii) Mix concentrated sulphuric acid (5 ml) and fuming nitric acid (3 ml) (CAUTION) and cool to room temperature. Add the organic compound (0·5 g) with continuous shaking and cooling. When the initial reaction has subsided, heat for 5 minutes on a boiling-water bath. Pour into cold water (50 ml) and filter off the solid product. Sometimes an oil is obtained, but this will usually solidify on vigorous stirring and scratching.

(iii) Treat the organic compound (0·5 g) dropwise with fuming nitric acid (3 ml) with continuous cooling in ice. When the reaction subsides, allow the mixture to stand at room temperature for 5 minutes before pouring it into water (50 ml). Filter off the solid obtained.

(iv) Dissolve the organic compound (0·5 g) in the minimum of glacial acetic acid and add a mixture of fuming nitric acid (2 ml) and glacial acetic acid (2 ml). Heat the mixture to boiling and allow to stand until cold. Pour into water (50 ml) and filter off the resulting solid.

(v) As described in (iv) above but keep the mixture at 20° by cooling in ice. After standing for 5 minutes, dilute with water and filter off the solid product.

In all the above cases, the crude product should be thoroughly washed with water and recrystallized from aqueous ethanol, ethanol or benzene.

(b) Alkyl 3,5-dinitrobenzoate

$$ROR + 3,5\text{-}(NO_2)_2C_6H_3\text{·}COCl \rightarrow 3,5\text{-}(NO_2)_2C_6H_3\text{·}CO\text{·}OR + RCl$$

Treat the alcohol-free ether (1 g) with powdered anhydrous zinc chloride (0·1 g) and 3,5-dinitrobenzoyl chloride (0·5 g). Reflux gently for 1 hour, pour the product into saturated aqueous sodium carbonate solution (10 ml) and heat on a boiling-water bath for 1 minute. Allow to cool and filter off the solid obtained. Wash with sodium carbonate solution and then water. Dry the solid and extract it with boiling carbon tetrachloride. Evaporate the excess of solvent and allow the derivative to crystallize.

(c) Picric acid complex

$$ArOR + 2,4,6\text{-}(NO_2)_3C_6H_2\text{·}OH \rightarrow [ArOR][2,4,6\text{-}(NO_2)_3C_6H_2\text{·}OH]$$

Prepare as described under *Amines, Primary and Secondary* (p. 26).

(d) Sulphonamide

Prepare as described under *Halides, Aryl* (p. 37).

(e) Bromination

where $n = 1, 2$ or 3.

Suspend or dissolve the ether (1 g) in glacial acetic acid, chloroform or carbon tetrachloride (5 ml) and add dropwise a solution of bromine in the same solvent until the colour of the bromine persists. Allow to stand for up to 15 minutes, adding more bromine solution if the colour fades. Evaporate the solvent (when using glacial acetic acid, pour into water) and recrystallize the product from ethanol.

HALIDES, ALKYL MONO-

(a) Thiouronium picrate

Treat the halide (1 ml) with a solution of thiourea (1·5 g) in water (4 ml) and ethanol (3 ml). Heat on a boiling-water bath until solution is complete and then for a further 15 minutes. Pour the resulting solution into an excess of 1% aqueous picric acid solution and filter off the precipitate which forms. Recrystallize from aqueous ethanol.

(b) 2-Naphthyl ether

$$AlkX + 2\text{-}C_{10}H_7O^- \rightarrow 2\text{-}C_{10}H_7 \cdot OAlk + X^-$$

Prepare a mixture of the halide (1 g), potassium hydroxide (1 g) and 2-naphthol (2 g) in ethanol (10 ml). Boil under reflux for 15 minutes and add a further portion of potassium hydroxide (2 g) in water (20 ml). Shake until a solid product is obtained, filter off and wash well with water. Recrystallize from aqueous ethanol. Isolate liquid products by ether extraction.

(c) 2-Naphthyl ether picrate

$$2\text{-}C_{10}H_7 \cdot OAlk + 2,4,6\text{-}(NO_2)_3C_6H_2 \cdot OH \rightarrow$$
$$[2\text{-}C_{10}H_7 \cdot OAlk][2,4,6\text{-}(NO_2)_3C_6H_2 \cdot OH]$$

To the product (0·5 g) from (b) above in ethanol (2 ml) add cold saturated alcoholic picric acid solution (5 ml). Collect the precipitate formed and recrystallize from ethanol.

(d) Oxidation of substituted benzyl halides to the corresponding carboxylic acids

$$ArCH_2X \xrightarrow{[O]} ArCO_2H$$

(i) *Chromic acid oxidation*

Carry out as described under *Alcohols* (p. 23).

(ii) *Potassium permanganate oxidation*

Mix the halide (1·5 g) with sodium hydroxide (1 g) and potassium permanganate (9 g) in water (100 ml). Reflux the solution until the colour of the permanganate is discharged and then filter off the manganese dioxide formed. Acidify the filtrate with concentrated hydrochloric acid and filter off the solid obtained. Recrystallize from water, aqueous ethanol or ethanol.

HALIDES, ALKYL POLY-

(a) Thiouronium picrate, e.g.

$$ClCH_2 \cdot CH_2Cl + 2H_2N \cdot CS \cdot NH_2 \xrightarrow{\text{picric acid}}$$

$$\left[\begin{matrix} H_2N \cdot C \cdot S \cdot CH_2 \cdot CH_2 \cdot S \cdot C \cdot NH_2 \\ \underset{NH_2}{\|} \qquad\qquad \underset{NH_2}{\|} \end{matrix} \right]^{2+} \quad 2[2,4,6\text{-}(NO_2)_3C_6H_2O]^-$$

Prepare as described under *Halides, Alkyl Mono-* (p. 35).

(b) **2-Naphthyl ether,** e.g.

$$CH_2Cl_2 + 2[2\text{-}C_{10}H_7O^-] \rightarrow [2\text{-}C_{10}H_7O]_2CH_2 + 2Cl^-$$

Prepare as described under *Halides, Alkyl Mono-* (p. 36).

HALIDES, ARYL

(a) **Sulphonamide**

$$ArX + ClSO_2 \cdot OH \underset{\text{excess}}{\longrightarrow} p\text{-}XAr \cdot SO_2Cl \xrightarrow{NH_3} p\text{-}XAr \cdot SO_2 \cdot NH_2$$

Prepare a solution of the halide (1 g) in dry chloroform (5 ml), cool in ice and add chlorosulphonic acid (3 ml). When the evolution of hydrogen chloride slackens, warm and maintain at room temperature for 30 minutes ($50°$ for 10 minutes if reaction is slow). Pour the product into crushed ice, separate the chloroform layer, dry over anhydrous sodium sulphate and evaporate the chloroform on a boiling-water bath. Add to the residue ammonia solution (0.88, 10 ml), boil for 10 minutes (fume cupboard), cool and dilute with water (10 ml). Filter off the crude sulphonamide and recrystallize from aqueous ethanol.

(b) **Nitro-derivative**

$$ArX \xrightarrow{NO_2^+} p\text{-}XAr \cdot NO_2$$

Prepare as described under *Ethers* (p. 33).

(c) **Picric acid complex**

$$ArX + 2,4,6\text{-}(NO_2)_3C_6H_2 \cdot OH \rightarrow [ArX][2,4,6\text{-}(NO_2)_3C_6H_2 \cdot OH]$$

Prepare as described under *Amines, Primary and Secondary* (picrate) (p. 27).

HYDRAZINE DERIVATIVES

(a) **Benzoyl derivative**

$$RNH \cdot NH_2 + C_6H_5 \cdot COCl \rightarrow RNH \cdot NH \cdot CO \cdot C_6H_5 + HCl$$

Prepare as described under *Amines, Primary and Secondary* (p. 26).

(b) **Hydrazone from hydrazine derivative**

$$RR'N \cdot NH_2 + C_6H_5 \cdot CHO \rightarrow RR'N \cdot N{:}CH \cdot C_6H_5 + H_2O$$

Dissolve the hydrazine derivative (0·5 g) in glacial acetic acid (1 ml) and water (1 ml). To this solution add benzaldehyde (0·5 g) and warm for 5 minutes. Add water (5 ml) and filter the solid obtained. Recrystallize from ethanol or glacial acetic acid.

HYDROCARBONS

(a) Diels-Alder adduct with maleic anhydride or benzoquinone, e.g.

Dissolve the hydrocarbon (1 g) in xylene (5–10 ml) and add powdered maleic anhydride (1 g) or benzoquinone (1 g). Reflux the mixture for 25 minutes and cool. If no solid separates, add small quantities of light petroleum (60–80°) to precipitate the product. Filter off the solid obtained and wash with a little light petroleum before recrystallizing from methanol, cyclohexane or xylene.

(b) Sulphonamide

$$ArH + ClSO_2{\cdot}OH \underset{\text{excess}}{\longrightarrow} ArSO_2Cl \xrightarrow{NH_3} ArSO_2{\cdot}NH_2$$

Prepare as described under *Halides, Aryl* (p. 37).

(c) Mercury derivative of alkynes

$$2RC{:}CH + K_2HgI_4 \rightarrow (RC{:}C)_2Hg$$

Dissolve mercuric chloride (6·6 g) in a solution of potassium iodide (16·3 g) in water (16·3 ml) and add 2 N sodium hydroxide (12·5 ml). Dissolve the alkyne (0·5 g) in ethanol (10 ml) and add it dropwise to the prepared solution (10 ml); filter off the precipitate immediately and wash with 50% aqueous ethanol. Recrystallize the product from ethanol or benzene.

(d) Picric acid and styphnic acid derivatives

Prepare as described under *Amines, Primary and Secondary* (picrate) (p. 27).

KETONES

(a) 2,4-Dinitrophenylhydrazone

$$RR'C{:}O + 2,4\text{-}(NO_2)_2C_6H_3{\cdot}NH{\cdot}NH_2 \rightarrow$$
$$2,4\text{-}(NO_2)_2C_6H_3{\cdot}NH{\cdot}N{:}CRR' + H_2O$$

Prepare as described under *Aldehydes* (p. 23).

(b) Semicarbazone

$$RR'C{:}O + H_2N{\cdot}NH{\cdot}CO{\cdot}NH_2 \rightarrow RR'C{:}N{\cdot}NH{\cdot}CO{\cdot}NH_2 + H_2O$$

Prepare as described under *Aldehydes* (p. 24).

(c) Oxime

$$RR'C{:}O + H_2N{\cdot}OH \rightarrow RR'C{:}N{\cdot}OH + H_2O$$

Dissolve the ketone (0·5 g) in ethanol (3 ml) and water (1 ml) and add hydroxylamine hydrochloride (0·3 g) followed by sodium hydroxide (0·5 g). When solution is complete, reflux for 5–10 minutes. Cool in ice and acidify with hydrochloric acid (use litmus paper). Filter off the product and recrystallize from ethanol.

(d) *p*-Nitrophenylhydrazone and phenylhydrazone

$$RR'C{:}O + ArNH{\cdot}NH_2 \rightarrow RR'C{:}N{\cdot}NHAr + H_2O$$

where $Ar = p\text{-}NO_2{\cdot}C_6H_4$— or C_6H_5—.
Prepare as described under *Aldehydes* (p. 24).

(e) Benzylidene derivative

$$
\begin{array}{c}
RCH_2 \\
| \\
C{:}O \\
| \\
R'CH_2
\end{array}
+ 2C_6H_5{\cdot}CHO \rightarrow
\begin{array}{c}
RC{:}CHC_6H_5 \\
| \\
C{:}O \\
| \\
R'C{:}CH{\cdot}C_6H_5
\end{array}
+ 2H_2O
$$

Shake a mixture of the ketone (0·5 g), a slight excess of benzaldehyde in ethanol and 4 N sodium hydroxide (0·5 ml). Allow to stand at room temperature until a crystalline product is obtained. Scratching the vessel with a glass rod will often induce crystallization. Filter off the product and recrystallize from ethanol.

NITRILES

(a) Amide

$$RCN \xrightarrow{H_2O_2} RCO \cdot NH_2$$

Prepare a solution containing 20 volume hydrogen peroxide (10 ml) and 2 N sodium hydroxide (2 ml); add the nitrile (0·5 g) and heat to 40° on a water bath. Shake frequently and finally filter off the solid product obtained. Wash and recrystallize from water, aqueous ethanol or ethanol.

(b) Nitro-derivative (for aromatic nitriles only)

$$ArCN \xrightarrow{NO_2^+} m\text{-}NO_2 \cdot Ar \cdot CN$$

Prepare as described under *Ethers*, Method (i) (p. 34).

(c) Hydrolysis to carboxylic acid

$$RCN \rightarrow RCO_2H$$

Reflux the nitrile (1 g) with either 8 N sodium hydroxide (5 ml) for aliphatic nitriles or 7 N sulphuric acid (5 ml) for aromatic nitriles for 1 hour. Cool the resulting solution and add excess hydrochloric acid (aliphatic nitriles) or water (aromatic nitriles). Characterize the aliphatic acid produced as its *p*-bromo-phenacyl ester as described under *Carboxylic Acids* (p. 31).

Aromatic acids produced may be filtered off and recrystallized from water, aqueous ethanol or ethanol.

NITRO-, HALOGENONITRO-COMPOUNDS AND NITRO-ETHERS

(a) Nitration

Prepare as described under *Ethers*, (p. 33).

(b) Benzylidene derivative

$$RCH_2 \cdot NO_2 + C_6H_5 \cdot CHO \rightarrow \underset{\overset{\|}{CH \cdot C_6H_5}}{RC \cdot NO_2} + H_2O$$

Prepare as described under *Ketones*, (p. 39).

(c) Oxidation of alkyl side-chain to carboxylic acid

$$ArCH_3 \xrightarrow{[O]} ArCO_2H$$

Prepare as described under *Alcohols*, (p. 23).

(d) Reduction to amine

$$RNO_2 \xrightarrow{[H]} RNH_2$$

Suspend the nitro-compound (1 g) in concentrated hydrochloric acid (10 ml) and add ethanol (2 ml) and tin (3 g). Cool until the initial reaction subsides and then heat under reflux for 30 minutes. Filter the solution, cool the filtrate and basify with 5 N sodium hydroxide, adding sufficient alkali to dissolve the precipitate of stannous hydroxide formed. Extract the free amine with ether, dry the ether extract over anhydrous sodium sulphate and evaporate the ether (CARE). Further conversion to crystalline derivatives should be carried out as described under *Amines, Primary and Secondary*, (p. 26).

(e) Partial reduction, e.g.

$$m\text{-}(NO_2)_2C_6H_4 \rightarrow m\text{-}NO_2 \cdot C_6H_4 \cdot NH_2$$

Dissolve the nitro-compound (1 g) in ethanol (10 ml) and add ammonia solution (0·88, 1 ml). Saturate the cold solution with hydrogen sulphide and reflux on a boiling-water bath for 30 minutes. Cool, resaturate with hydrogen sulphide and reflux for a further 30 minutes. Pour into cold water and filter the solid obtained. Extract the solid with 2 N hydrochloric acid, basify this extract with ammonia solution (0·88) and filter the resulting nitro-amine. Recrystallize from aqueous ethanol, ethanol or benzene.

PHENOLS

(a) Acetate

$$ArOH + CH_3 \cdot COCl \rightarrow ArO \cdot CO \cdot CH_3 + HCl$$

Dissolve the phenol (0·5 g) in dry pyridine (0·5 ml) and add acetyl chloride (0·5 ml) dropwise. Shake well after each addition and cool if the temperature rises rapidly. When the addition of acetyl chloride is complete, heat to 50–60° for 5 minutes. Cool, pour into water (15 ml) and stir until a solid is obtained. Filter off the solid and recrystallize from ethanol or aqueous ethanol.

(b) Benzoate and toluene-*p*-sulphonate

$$ArOH + C_6H_5 \cdot COCl \rightarrow ArO \cdot CO \cdot C_6H_5 + HCl$$

$$ArOH + p\text{-}CH_3 \cdot C_6H_4 \cdot SO_2Cl \rightarrow p\text{-}CH_3 \cdot C_6H_4 \cdot SO_2 \cdot OAr + HCl$$

41

4

Prepare as described under *Alcohols* (p. 22). For nitrophenols it is preferable to replace this Schotten-Baumann method by one in which pyridine is the base, as in (c) below.

(c) *p*-Nitrobenzoate and 3,5-dinitrobenzoate

$$ArOH + Ar'COCl \rightarrow ArO \cdot COAr' + HCl$$

where $Ar' = p\text{-}NO_2 \cdot C_6H_4^-$ or $3,5\text{-}(NO_2)_2C_6H_3^-$.
Prepare as described under *Alcohols*, (p. 22).

(d) Aryloxyacetic acid

$$ArOH + ClCH_2 \cdot CO_2H \rightarrow ArO \cdot CH_2 \cdot CO_2H + HCl$$

To a solution of the phenol (0·5 g) in 5 N sodium hydroxide (3 ml) add chloroacetic acid (0·5 g) (CAUTION: this acid must not be allowed to come into contact with the skin). Add a little water if any solid is formed in the hot solution. Heat on a boiling-water bath for 1 hour, cool, acidify with 2 N hydrochloric acid to Congo Red and extract with ether. Extract the ethereal layer with 2 N sodium carbonate solution. If the sodium salt of the acid separates, remove by filtration and treat the solid with 2 N hydrochloric acid. The resulting solid is the required derivative. If no solid separates, acidify the sodium carbonate extract with 2 N hydrochloric acid. Filter off the solid obtained. Recrystallize the product from water, aqueous ethanol or ethanol.

(e) Derivative for picric and styphnic acids, e.g.

$$2,4,6\text{-}(NO_2)_3C_6H_2 \cdot OH + C_{10}H_8 \rightarrow [2,4,6\text{-}(NO_2)_3C_6H_2 \cdot OH][C_{10}H_8]$$

Prepare saturated solutions of either picric or styphnic acid and naphthalene in ethanol and mix. Warm gently for a few minutes and cool. A crystalline product is readily obtained which may be recrystallized from ethanol.

QUINONES

(a) Oxime, e.g.

Prepare as described under *Ketones*, (p. 39).

(b) **Semicarbazone,** e.g.

Prepare as described under *Aldehydes*, (p. 24).

(c) **Quinol,** e.g.

Dissolve or suspend the quinone (1 g) in benzene (5–10 ml) and treat with a solution (20 ml) of sodium hydrosulphite (10%) in N sodium hydroxide. Shake until the quinone colour has disappeared and separate the aqueous layer. Cool this (ice) and acidify with concentrated hydrochloric acid. Filter off the solid obtained and recrystallize from water or ethanol.

SULPHONIC ACIDS AND THEIR DERIVATIVES

(a) **Amide**

$$RSO_2 \cdot OH + PCl_5 \rightarrow RSO_2Cl \xrightarrow{NH_3} RSO_2 \cdot NH_2$$

Mix the dry sulphonic acid (1 g) or the dry sodium salt (1 g) with phosphorus pentachloride (2 g) and heat on a boiling-water bath, taking care to exclude water vapour from the reaction vessel. When reaction has ceased, add water (15 ml) and stir. Decant the water and add ammonia solution (0·88, 3 ml) to the residue. Heat on a boiling-water bath for 5–10 minutes and cool. Filter off the solid product, wash well with water and recrystallize from water or aqueous ethanol.

(b) Anilide

$$RSO_2 \cdot OH + PCl_5 \rightarrow RSO_2Cl \xrightarrow{C_6H_5 \cdot NH_2} RSO_2 \cdot NH \cdot C_6H_5$$

Use the method described above but replace the ammonia solution with aniline (1 ml).

(c) S-Benzylisothiouronium salt

$$RSO_2 \cdot O^- Na^+ + \begin{bmatrix} C_6H_5 \cdot CH_2 \cdot S \cdot C:NH_2 \\ | \\ NH_2 \end{bmatrix}^+ Cl^- \rightarrow$$

$$\begin{bmatrix} C_6H_5 \cdot CH_2 \cdot S \cdot C:NH_2 \\ | \\ NH_2 \end{bmatrix}^+ [O \cdot SO_2 R]^- + NaCl$$

Prepare the sodium salt of the sulphonic acid (0·5 g) in water (3 ml) by addition of 2 N sodium hydroxide until the solution is just alkaline to phenolphthalein. Neutralize the excess of alkali with a further addition of the sulphonic acid (or 2 N hydrochloric acid) and add a solution of S-benzylthiouronium chloride (2 g) in water (5 ml). Cool in ice, filter off the crystalline product and recrystallize from water, aqueous ethanol or ethanol.

(d) Xanthyl derivative of sulphonamides

Treat the sulphonamide (0·5 g) with xanthydrol (0·5 g) in glacial acetic acid (25 ml) and heat the mixture until solution is complete. Allow to stand at room temperature until a solid separates. Water may be added if no solid separates. Filter off the product and recrystallize from aqueous dioxan.

(e) Benzoyl derivative of sulphonamides

$$RSO_2 \cdot NH_2 + C_6H_5 \cdot COCl \rightarrow RSO_2 \cdot NH \cdot CO \cdot C_6H_5 + HCl$$

Prepare as described under *Alcohols*, (p. 22).

(f) Acetyl derivative of sulphonamides

$$RSO_2 \cdot NH_2 + CH_3 \cdot COCl \rightarrow RSO_2 \cdot NH \cdot CO \cdot CH_3 + HCl$$

To the sulphonamide (1 g) add acetyl chloride (3 ml) and reflux for 30 minutes, adding glacial acetic acid (up to 2 ml) if solution is not complete. Remove the excess of acetyl chloride by vacuum distillation and pour the residue into ice cold water (25 ml). Stir the product until it solidifies, filter off, wash well with water and recrystallize from aqueous ethanol.

THIOETHERS (SULPHIDES)

(a) Sulphone

$$RSR' \xrightarrow{[O]} RSO_2R'$$

Dissolve the thioether (1 g) in the minimum of glacial acetic acid and add 3% potassium permanganate solution as long as the colour is discharged. If starting material is precipitated during this addition, more glacial acetic acid should be added. When reaction is complete, pass in sulphur dioxide until the manganese dioxide precipitate has just dissolved. Add crushed ice and filter off the solid sulphone. Wash well with water and recrystallize from ethanol.

THIOLS AND THIOPHENOLS

(a) 2,4-Dinitrophenyl sulphide

$$RSH + 2,4\text{-}(NO_2)_2C_6H_3Cl \rightarrow 2,4\text{-}(NO_2)_2C_6H_3 \cdot SR + HCl$$

Dissolve the thiol (1 g) in ethanol (30 ml) and add sodium hydroxide (0·4 g) in ethanol (3 ml) followed by 2,4-dinitrochlorobenzene (2 g) (CAUTION—skin irritant) in ethanol (10 ml). Reflux on a boiling-water bath for 10 minutes, filter and allow the filtrate to cool. Recrystallize the product from ethanol.

(b) Hydrogen 3-nitrophthaloyl derivative

Prepare as described under *Alcohols* (p. 23).

(c) 3,5-Dinitrobenzoyl derivative

$$RSH + 3,5\text{-}(NO_2)_2C_6H_3 \cdot COCl \rightarrow 3,5\text{-}(NO_2)_2C_6H_3 \cdot CO \cdot SR + HCl$$

Prepare as described under *Alcohols* (p. 22).

45

TABLES OF ORGANIC COMPOUNDS
AND THEIR DERIVATIVES

EXPLANATORY NOTES ON THE TABLES
OF COMPOUNDS AND DERIVATIVES

1. In each table the compounds are listed in the order of increasing boiling point if they are liquids or solids melting below 40°. Compounds which melt at 40° or above are divided from the liquids by a horizontal line and are arranged in the order of increasing melting point although the boiling point is sometimes also included.

2. Boiling points are given at atmospheric pressure except for a few high boiling compounds whose boiling points are given at reduced pressure and are written thus: 94/12 mm, which means a boiling point of 94° at 12 mm pressure.

3. Boiling and melting points are given to the nearest whole number and as one value only. This approximation is made for simplicity and because of slight variation in the degree of accuracy of different thermometers and in the personal element in determining the melting or boiling point. For example, a melting point which is recorded in the literature as 172–174° or as 173·5° is given in the tables as 173°.

4. The figures given in the columns of derivatives are melting points. Two different values of a boiling or melting point are recorded in the chemical literature for a few compounds. The second (usually less frequently encountered) value is given in the tables in parentheses below the more common value.

5. For compounds which exist in enantiomorphic forms, the constants given are those of the racemic or (±)-modification unless otherwise stated.

6. The following abbreviations are used in the tables:

anhyd. anhydrous	*dil*. dilute
aq. aqueous	*hyd*. hydrate
conc. concentrated	*insol*. insoluble
d. decomposition	*sol*. soluble
deriv. derivative	*subl*. sublimes

7. When the colour of a compound is other than white it is given in the last column of the table which also contains supplementary information which may assist in the identification of the compound.

TABLES OF ORGANIC COMPOUNDS AND THEIR DERIVATIVES

Table 1. Acetals

Acetal	B.p.	Aldehyde	p-Nitrophenyl-hydrazone (p. 24)	2,4-Dinitro-phenylhyd-razone (p. 23)	Alcohol	K alkyl xanthate (p. 22)
Dimethoxymethane (Methylal)	45	Formaldehyde	182	167	Methanol	182
1,1-Dimethoxyethane (Dimethylacetal)	64	Acetaldehyde	128	168	Methanol	182
Diethoxymethane (Ethylal)	89	Formaldehyde	182	167	Ethanol	225
1,1-Diethoxyethane (Acetal)	102	Acetaldehyde	128	168	Ethanol	225
1,1-Diethoxyprop-2-ene (Acrolein acetal)	126	Propenal (Acrolein)	151	165	Ethanol	225
1,1-Dipropoxymethane	140	Formaldehyde	182	167	Propan-1-ol	233
1,1-Dibutoxyethane	187	Acetaldehyde	128	168	Butan-1-ol	255
αα-Dimethoxytoluene	198	Benzaldehyde	192	237	Methanol	182
αα-Diethoxytoluene	222	Benzaldehyde	192	237	Ethanol	225

Table 2. Alcohols (C, H and O)

	B.p.	M.p.	3,5-Di-nitro-benzo-ate (p. 22)	H 3-nitro-phthal-ate (p. 23)	p-Nitro-benzo-ate (p. 22)	Notes
Methanol	65		109	153*	96	* Anhydrous; monohy-drate melts <100 but if dried at 80, it becomes an-hydrous.
Ethanol	78		94	157	56	Gives iodoform test.
Isopropanol	83		122	154	110	Gives iodoform test.
t-Butanol	83	25	142		116	
Propanol	97		75	144	35	
Prop-2-enol (Allyl alcohol)	97		50	124	28	Unsaturated.
Butan-2-ol (s-Butanol)	100		76	131	25	Gives iodoform test.
2-Methylbutan-2-ol (t-Pentyl alcohol)	102		118		85	
2-Methylpropan-1-ol (Isobutanol)	108		88	183	69	
(±)-3-Methylbutan-2-ol	113		76	127		Gives iodoform test.
Pentan-3-ol	116		100	121	17	

47

Table 2 (cont). Alcohols (C, H and O)

	B.p.	M.p.	3,5-Di-nitro-benzo-ate (p. 22)	H 3-nitro-phthal-ate (p. 23)	p-Nitro-benzo-ate (p. 22)	Notes
Butan-1-ol	117		64	147	35	
Pentan-2-ol	120		62	103	17	Gives iodoform test.
3-Methylpentan-3-ol	123		96		69	
2-Methoxyethanol (Methyl cellosolve)	125			129	50	
2-Methylbutan-1-ol (Active pentyl or amyl alcohol)	128		70	158		
3-Methylpentan-1-ol (Isopentyl alcohol)	132		62	164	21	
4-Methylpentan-2-ol	132		65	166	26	Gives iodoform test.
2-Ethoxyethanol (Ethyl cellosolve)	135		75	121*		* Anhydrous; monohydrate, 94.
Hexan-3-ol	136		77	127		
Pentan-1-ol	138		46	136	oil	
2,4-Dimethylpentan-3-ol	140		38	151	40	
Cyclopentanol	140		115		62	
3-Hydroxybutan-2-one (Acetoin)	145					See Table 26. Gives iodoform test.
1-Hydroxypropan-2-one (Acetol)	146					See Table 26.
2-Methylpentan-1-ol	148		50	145		
2-Ethylbutan-1-ol	149		51	147		
4-Methylpentan-1-ol	152		70	140		
Hexan-1-ol	156		61	124		
Heptan-2-ol	158		50			Gives iodoform test. $K_2Cr_2O_7$—H_2SO_4 (p. 23) → heptan-2-one.
Cyclohexanol	161	25	113	160	52	
Furfuryl alcohol	170		81		76	
Heptan-1-ol	176		48	127	oil	
Tetrahydrofurfuryl alcohol	177		83		47	
Octan-2-ol	179		32		28	Gives iodoform test. $K_2Cr_2O_7$—H_2SO_4 (p. 23) → octan-2-one.
2-Ethylhexan-1-ol	184			108		
Propane-1,2-diol (Propylene glycol)	187		147		127	
3,5,5-Trimethylhexan-1-ol	193		62	150		
Octan-1-ol	194		62	128	oil	
(—)-Linalyl alcohol ((—)-Linaloöl)	197				70	Unsaturated

Table 2 (cont). Alcohols (C, H and O)

	B.p.	M.p.	3,5-Di-nitro-benzo-ate (p. 22)	H 3-nitro-phthal-ate (p. 23)	p-Nitro-benzo-ate (p. 22)	Notes
Ethane-1,2-diol (Ethylene glycol)	197		169		140	Ditoluene-p-sulphonate, 93 (p. 22).
1-Phenylethanol	202	20	94		43	
Benzyl alcohol	205		113	183	85	
Nonan-1-ol	214		52	125		
Propane-1,3-diol (Trimethylene glycol)	214		178		119	Ditoluene-p-sulphonate, 93 (p. 22).
Isoborneol	216		138		129	
2-Phenylethanol	219		108	123	62	
α-Terpineol	221	35	78		139	
Tetradecan-1-ol (Myristic alcohol)	221	39	67	123	51	
Geraniol	229		63	117	35	Unsaturated; Br$_2$ → tetra-bromide, 70.
Butane-1,4-diol	230	19			175	Dibenzoate, 81; ditoluene-p-sulphonate, 94 (p. 22).
Decan-1-ol	231	6	57	123	30	
2-Phenoxyethanol (Ethylene glycol monophenyl ether)	237 (245)			113	63	Toluene-p-sulphonate, 80 (p. 22).
3-Phenylpropan-1-ol (Hydrocinnamyl alcohol)	237		92	117	46	
Undecan-1-ol	243	15	55	123	30	
Di-(2-hydroxyethyl) ether (Diethylene glycol)	244		150			
Cinnamyl alcohol	257	33	121		78	Unsaturated; Br$_2$ → di-bromide, 74.
Dodecan-1-ol (Lauryl alcohol)	259	25	60	124	43	
Glycerol	290d	18			188	
(—)-Menthol	216	42	153		61	
Hexadecan-1-ol (Cetyl alcohol)		50	66	122	52 (58)	
Heptadecan-1-ol		54	121	121	53	
But-2-yn-1,4-diol		55	190			Dibenzoate, 76; di-p-tolu-enesulphonate, 94 (p. 22).
Octadecan-1-ol (Stearyl alcohol)		59	66	119	64	
Diphenylmethanol (Benzhydrol)		69	141		131	
D-Glucitol (D-Sorbitol)		111*				* Anhydrous; hydrate, 90. Hexa-acetate, 99; hexa-benzoate, 129 (p. 22).

Table 2 (cont). Alcohols (C, H and O)

	B.p.	M.p.	3,5-Di-nitro-benzo-ate (p. 22)	H 3-nitro-phthal-ate (p. 23)	p-Nitro-benzo-ate (p. 22)	Notes
Benzoin		133			123	See *Table 26.*
Furoin		136				See *Table 26.*
Triphenylmethanol		162				Acetate, 87; benzoate, 162 (p. 22).
D-Mannitol		166				Hexa-acetate, 126; hexa-benzoate, 148 (129) (p. 22).
D-Galactitol (Dulcitol)		188				Hexa-acetate, 171; hexa-benzoate, 188 (p. 22).
(+)-Borneol	212	208	154		153	
Meso-inositol		225	86			Hexa-acetate, 212 subl.; hexabenzoate, 258 (p. 22).
Pentaerythritol		262 (253)				Tetra-acetate, 84; tetra-benzoate, 99 (p. 22).

Table 3. Alcohols (C, H, O and halogen or N)

	B.p.	M.p.	3,5-Di-nitro-benzo-ate (p. 22)	H 3-nitro-phthal-ate (p. 23)	p-Nitro-benzo-ate (p. 22)	Notes
1-Chloropropan-2-ol	127		77			Gives iodoform test.
2-Chloroethanol (Ethylene chlorohydrin)	129		95 (88)	98	56	
2-Chloropropan-1-ol	133		76			
2-Dimethylaminoethanol	135					See *Table 12.*
2-Bromoethanol	149d		85	172		
2,2,2-Trichloroethanol	151	19	142		71	
2-Diethylaminoethanol	161					See *Table 12.*
3-Chloropropan-1-ol	161		77			
(±)-2-Hydroxypropylamine (Isopropanolamine)	163					Picrate, 142; see also *Table 8.*
2-Aminoethanol	171					Picrate, 159; reacts with phthalic anhydride → β-hydroxyethylimide, 127.
m-Nitrobenzyl alcohol	175/3 mm	27				Benzoate, 71; $K_2Cr_2O_7$—H_2SO_4 → acid, 140 (p. 23).
o-Nitrobenzyl alcohol		74				Benzoate, 101; $K_2Cr_2O_7$—H_2SO_4 → acid, 146 (p. 23).
p-Nitrobenzyl alcohol		93				Benzoate, 94; $K_2Cr_2O_7$—H_2SO_4 → acid, 240 (p. 23).

Table 4. Aldehydes (C, H and O)

	B.p.	M.p.	2,4-Dinitro-phenyl-hydraz-one (p. 23)	Semi-carb-azone (p. 24)	Di-meth-one (p. 25)	p-Nit-rophen-ylhyd-razone (p. 24)	Notes
Formaldehyde	−21		167	169d	189	182	40% aq. solution is formalin.
Acetaldehyde	20		168	163	140	128	Gives iodoform test.
Propionaldehyde	50		155	154*	155	124	* Recryst. from water.
Glyoxal	50		327	270	186* mono	310d	* Di-, 228
Propenal (Acrolein)	52		165	171	192	151	Unsaturated. Gives iodoform test.
2-Methylpropionalde-hyde (Isobutyraldehyde)	64		182	125	154	131	
2-Methylprop-2-enal (α-Methylacrolein)	73		206	198			Unsaturated.
Butyraldehyde	74		125	105	136	91	
2,2-Dimethylpropion-aldehyde (Pivalaldehyde)	75		209	190		119	
2-Methylbutyraldehyde (Isovaleraldehyde)	92		123	132 (107)	155	110	
2-Methylbutyraldehyde	93		121	103			
Pentanal (Valeraldehyde)	103		107		105	74	
But-2-enal (Crotonaldehyde)	103		196	200*	186	184	* Varies with rate of heating. Unsaturated.
5-Hydroxymethylfur-furaldehyde	114	35	184	195d (166)		185	
2-Ethylbutyraldehyde	116		134	98	102		
Paraldehyde	124						Gives acetaldehyde on warming with trace of conc. sulphuric acid.
Hexanal (Caproaldehyde)	129		104	108	109	80	
3-Methylbut-2-enal (β-Methylcrotonalde-hyde)	135		182	221			Unsaturated.
Tetrahydrofurfural	144		134	166	123		
Heptanal (Heptaldehyde)	156		108	109	103 (135)	73	
Furfural	161		202*	203	160d	154	* Variable. Gives iodoform test.
Hexahydrobenzaldehyde	162		172	174			

Table 4 (cont). Aldehydes (C, H and O)

	B.p.	M.p.	2,4-Di-nitro-phenyl-hydraz-one (p. 23)	Semi-carb-azone (p. 24)	Di-meth-one (p. 25)	p-Nit-rophen-ylhyd-razone (p. 24)	Notes
2-Ethylhexanal	163		121	254d			
Succinaldehyde	169		280				Oxime, 172 (p. 24).
Octanal	171		106	98	90	80	
(Caprylaldehyde)							
Benzaldehyde	179		237	222*	195	192	* Varies with rate of heating. Smell of bitter almonds.
Nonanal	185		100	100	86		
(Pelargonaldehyde)							
5-Methylfurfural	187		212	211		130	
Phenylacetaldehyde	194	33	121	156	165	151	
Salicylaldehyde	196		248*	231	211	227	* From ethanol. See also *Table 30*.
m-Tolualdehyde	199		194	224	172	157	
o-Tolualdehyde	200		194	212	167	222	
p-Tolualdehyde	204		234	234		200	
(+)-Citronellal	206		77	83	78		Unsaturated.
Decanal	208		104	102	92		
(Capraldehyde)							
Phenoxyacetaldehyde	215d		130	145			Oxime, 95 (p. 24).
3-Phenylpropional-dehyde	224		149	127		123	
(Hydrocinnamaldehyde)							
Geranial	228d		116 (108)	164		195	Unsaturated.
(Citral a)							
m-Methoxybenzaldehyde	230		218	233d		171	
p-Isopropylbenzaldehyde	235		244	211	171	190	
(Cuminaldehyde)							
o-Methoxybenzaldehyde	245	38	253	215	188	205	
p-Methoxybenzaldehyde	248		253d	210	145	160	
(Anisaldehyde)							
Cinnamaldehyde	252		255d	215	219 (213)	195	Unsaturated.
3,4-Methylenedioxy-benzaldehyde	263	37	266d	234	178	200	
(Piperonal)							
1-Naphthaldehyde	292	34		221		234	
(α-Naphthaldehyde)							
Tetradecanal	155/10 mm	23	108	107		95	
(Myristaldehyde)							
Hexadecanal	200/29 mm	34	108	108		97	
(Palmitaldehyde)							
Octadecanal	212/22 mm	38	110	108		101	
(Stearaldehyde)							

Table 4 (cont). Aldehydes (C, H and O)

	B.p.	M.p.	2,4-Di-nitro-phenyl-hydraz-one (p. 23)	Semi-carb-azone (p. 14)	Di-meth-one (p. 25)	p-Nit-rophen-ylhyd-razone (p. 24)	Notes
Dodecanal (Lauraldehyde)		44	106	106		90	
2,3-Dimethoxybenz-aldehyde		54	223 (264)	231	150		
Phthalaldehyde		56					Phenylhydrazone, 191 (p. 24).
3,4-Dimethoxybenzaldehyde (Veratraldehyde)		58	264	177	173		
2-Naphthaldehyde (β-Naphthaldehyde)		60	270	245		230	
2,4-Dimethoxybenzaldehyde		69	257				Oxime, 106 (p. 24).
4-Hydroxy-3-methoxy-benzaldehyde (Vanillin)		80	271d	240d	197	227	Bisulphite addition compound is sol. in water.
Phenylglyoxal, hydrate		91		217d		310	Dioxime, 168; mono-oxime, 128 (p. 24).
m-Hydroxybenzaldehyde	104		260	199		222	See also *Table 30*.
Terephthalaldehyde	116			225d		281*	* Sinters at 272. Oxime, 200 (p. 24).
p-Hydroxybenzaldehyde	117		280d	222	189	266	See also *Table 30*; bisul-phite compound is sol. in water.
2,4-Dihydroxybenzaldehyde (β-Resorcylaldehyde)	135		286d (302)	260d	226		See also *Table 30*.
(±)-Glyceraldehyde (dimer)	142		167	160d	197		
3,4-Dihydroxybenzaldehyde (Protocatechualdehyde)	154		275d	230d	143d		See also *Table 30*.
3,5-Dihydroxybenzaldehyde	157			223		280*	* Chars without melting.

Table 5. Aldehydes (C, H, O and halogen or N)

	B.p.	M.p.	2,4-Dinitrophenylhydrazone (p. 23)	Semicarbazone (p. 24)	Dimethone (p. 25)	p-Nitrophenylhydrazone (p. 24)	Notes
Trichloroacetaldehyde (Chloral)	98		131	90d			
Tribromoacetaldehyde (Bromal)	174*						* Anhydrous; hydrate, 54. Oxime, 115 (p. 24).
o-Iodobenzaldehyde	206	37	215	206			
o-Chlorobenzaldehyde	208	11	208	225	205	249	
m-Chlorobenzaldehyde	213	18	255	228		216	
o-Bromobenzaldehyde	230	22		214		240	
m-Bromobenzaldehyde	234			205		220	
o-Aminobenzaldehyde		40		247		219	Oxime, 135 (p. 24).
p-Diethylaminobenzaldehyde		41	206	214			Oxime, 93 (p. 24).
o-Nitrobenzaldehyde		44	250d	256		263	
p-Chlorobenzaldehyde	214	47	268	231		239	
Trichloroacetaldehyde hydrate (Chloral hydrate)		57	131	90d			
p-Bromobenzaldehyde		57	257* 128*	228		208	* Polymorphs
m-Iodobenzaldehyde		57	258	226		213	
m-Nitrobenzaldehyde		58	293d	246	198	247	
2,4-Dichlorobenzaldehyde		71	226			256	Oxime, 136 (p. 24).
p-Dimethylaminobenzaldehyde		74	237	222		182	
p-Iodobenzaldehyde		78	257	224		201	
p-Nitrobenzaldehyde		106	320	220	190	249	
5-Bromosalicylaldehyde		106		297d			Oxime, 126 (p. 24). See also *Table 31.*

Table 6. Amides (primary), Imides, Ureas, Thioureas and Guanidines

	M.p.	Xanthyl deriv. (p. 25)	Carb-oxylic acid (p. 25)	Notes
Formamide (b.p. 193)	3	184		
Ethyl carbamate (Urethane)	49	169		
Methyl carbamate	54	193		
Propynamide (Propiolic amide)	61			Unsaturated.
Propionamide	81	214		
Acetamide	82	240		
Propenamide (Acrylamide)	84			Unsaturated.
2-Phenylpropionamide	92	158		
Maleimide	93		130	Unsaturated.
Heptanamide	96	154		
Semicarbazide	96			Acetaldehyde semicarbazone, 163.
Hexanamide	100	160		
N-Methylurea	101	230		Acetyl deriv., 180.
3-Phenylpropionamide	105	189	48	
Hexadecanamide (Palmitamide)	106	141	62	
Pentanamide (Valeramide)	106	167		
Octanamide	106	148	16	
Octadecanamide (Stearamide)	108	141	70	
Butyramide	116	186		
Chloroacetamide	120	209	*	* Hydrolysis gives hydroxyacetic acid, 80.
Cyanacetamide	123	223		
Succinimide	125	246	185	
Benzamide	128	223	122	
2-Methylpropionamide (Isobutyramide)	128	211		
Urea	132	274		
3-Methylbutyramide (Isovaleramide)	135	183		
Salicylamide	139		158	See also Table 31.
2-Furamide	142	209	133	
o-Toluamide	142	200	105	
N-Phenylurea	147	225		
N-Phenylthiourea	154			
Benzilamide	155		150	
Phenylacetamide	157	196	76	
p-Toluamide	160	225	180	
Malonamide	170	270	133d	
Guanidine hydrochloride	172			Cold conc. $HNO_3 + H_2SO_4 \rightarrow$ nitroguanidine, 230d.

Table 6 (cont). Amides (primary), Imides, Ureas, Thioureas and Guanidines

	M.p.	Xanthyl deriv. (p. 25)	Carboxylic acid (p. 25)	Notes
p-Ethoxyphenylurea (Dulcin)	173			Heating above m.p. → di-(p-ethoxyphenyl)urea, 235.
Thiourea	180			Heating at m.p. → NH_4CNS which with aq. $FeCl_3$ gives red colour.
NN-Dimethylurea	182	250		
Thiosemicarbazide	182			Acetyl deriv., 165; benzaldehyde thiosemicarbazone, 160.
1,1-Diphenylurea (Carbanilide)	189	180		
o-Tolylurea	191	228		
Biuret*	192d	260		* $H_2N\cdot CO\cdot NH\cdot CO\cdot NH_2$. With trace of $CuSO_4$ + dil. NaOH → red colour; an excess of $CuSO_4$ gives violet colour.
Guanidine carbonate	197			Cold conc. HNO_3 + H_2SO_4 → nitroguanidine, 230d.
p-Nitrobenzamide	201	232	240	
4-Nitrophthalimide	202		165	
N-Cyanoguanidine (Dicyandiamide)	208			
Guanidine nitrate	214			Cold conc. HNO_3 + H_2SO_4 → nitroguanidine, 230d.
3-Nitrophthalimide	218		218	
Phthalamide	220		200d	Loses NH_3 near its m.p. to give phthalimide, 233.
o-Benzoic sulphimide (Saccharin)	230	199		
Phthalimide	233	177	200d	
Succinamide	260d	275	125	

Table 7. Amides, *N*-substituted

(*N*-Substituted amides listed as acetyl and benzoyl derivatives of amines and amino-acids in *Tables 8–11* and *13* should be consulted together with the following list). Hydrolysis (p. 26) should be followed by identification of the acid and amine.

	B.p.	Foot-note		M.p.	Foot-note
NN-Dimethylformamide	153		*N*-Chlorosuccinimide	150	
NN-Dimethylacetamide	165		*N*-*p*-Tolylacetamide	153	d
NN-Diethylformamide	176		*N*-Methyl-*N*-(*p*-nitrophenyl)-acetamide	153	
	M.p.		*N*-Phenylsuccinimide	156	
N-Phenylformamide (Formanilide)	46		(Succinanil)		
3-Oxobutyramide (Acetoacetamide)	54		*N*-*p*-Tolylbenzamide	158	
N-Ethyl-*N*-phenylacetamide (*N*-Ethylacetanilide)	54	a	*N*-1-Naphthylacetamide	160	
N-Phenyloctanamide (Caprylanilide)	57		*N*-1-Naphthylbenzamide	161	
N-Phenylpentanamide (Valeranilide)	63		*N*-Phenylbenzamide	163	e
N-Phenyl-3-oxobutyramide (Acetoacetanilide)	85		*N*-*p*-Bromophenylacetamide (*p*-Bromoacetanilide)	167	
N-Phenyloctadecanamide (Stearanilide)	94		*N*-Bromosuccinimide	174	
N-Benzylsuccinimide	99		*N*-*p*-Chlorophenylacetamide	178	
NN-Diphenylacetamide	101		*NN'*-Diacetyl-*o*-phenylenedi-amine	186	
N-Methyl-*N*-phenylacetamide	102		*N*-Benzoylglycine (Hippuric acid)	187	
N-Phenylpropionamide	106	b	*NN'*-Diacetyl-*m*-phenylenedi-amine	190	
N-Phenylacetamide (Acetanilide)	114		*N*-Phenyl-*p*-chlorobenzamide	194	
N-Benzylphthalimide	116		*N*-Phenylphthalimide	205	
N-Phenyl-*o*-chlorobenzamide	118		*N*-(*p*-Hydroxyphenyl)-*N*-methyl-acetamide	240	
N-Phenyl-*o*-toluamide	125		Barbituric acid	245	
N-Phenyl-*m*-toluamide	126		*NN'*-Diacetyl-*p*-phenylenedi-amine	303	
N-(*p*-Methoxyphenyl)acetamide	127				
N-(*o*-Methoxyphenyl)acetamide	129				
N-(2,4-Dimethylphenyl)-acetamide	130				
N-Phenylbromoacetamide	131				
N-(*p*-Ethoxyphenyl)acetamide (Phenacetin)	135	c			
N-(*o*-Tolyl)benzamide	142				
N-Phenyl-*p*-toluamide	144				

Footnotes:

a. Conc. $HNO_3 + H_2SO_4$ at 40° → *p*-nitro deriv., 118.

b. Conc. $HNO_3 + H_2SO_4$ at 0° → *p*-nitro deriv., 182.

c. Warm 10% HNO_3 → *m*-nitro deriv., 103.

d. Br_2-acetic acid → *m*-bromo deriv., 117.

e. Br_2-acetic acid → *p*-bromo deriv., 202.

57

5

Table 8. Amines, primary aliphatic

	B.p.	M.p.	Acetyl deriv. (p. 26)	Benzoyl deriv. (p. 26)	Toluene-p-sulphonyl deriv. (p. 26)	2,4-Dinitrophenyl deriv. (p. 27)	Notes
Methylamine	−7		28	80	77	178 ⎫	Normally supplied
Ethylamine	17			69	62	113 ⎬	as an aqueous
Isopropylamine	32			100	51	95 ⎭	solution.
1,1-Dimethylethylamine (t-Butylamine)	46			134			Picrate, 198 (p. 27).
Propylamine	49			85	52	97	
Prop-2-enylamine (Allylamine)	58			Oil	64	76	Picrate, 140 (p. 27).
(±)-1-Methylpropyl- amine (s-Butylamine)	63			76	55		Picrate, 140 (p. 27).
2-Methylpropylamine (Isobutylamine)	69			57	78	94 (80)	
Butylamine	77			42	65	90	Picrate, 151 (p. 27).
3-Methylbutylamine (Isopentylamine)	96				65	91	Picrate, 138 (p. 27).
Pentylamine	104					81	Picrate; 139 (p. 27).
Ethylenediamine	116	8	172	249	160	306	
(±)-Propane-1,2-diamine (Propylenediamine)	119		139	192	103		
Hexylamine	130			40		39	Picrate, 127; benzenesul- phonamide, 96 (p. 26).
Cyclohexylamine	134		104	147	87	156	
Propane-1,3-diamine (Trimethylenediamine)	136		126	147	148		
Butane-1,4-diamine (Tetramethylenediamine)	159	27	137	177	224		
(±)-2-Hydroxypropyl- amine (Isopropanolamine)	163						Picrate, 142; see also *Table 3.*
2-Aminoethanol	171				Oil		Picrate, 159; see also *Table 3.*
Benzylamine	184			60	106	116	116
(±)-1-Phenylethylamine	185			57	120		118
3-Aminopropanol	188						Picrate, 222; see also *Table 3.*
2-Phenylethylamine	197			51	116	64	154
(−)-Menthylamine	205			145	156		Picrate, 215 (p. 27).
Dodecylamine	247	27				73	
Diphenylmethylamine (Benzhydrylamine)	303			146	172		
Hexane-1,6-diamine (Hexamethylenediamine)	204	42			155		Picrate, 220 (p. 27).

Table 9. Amines, primary aromatic (C, H, (O) and N)

	B.p.	M.p.	Acetyl deriv. (p. 26)	Benz-oyl deriv. (p. 26)	Tolu-ene-p-sulph-onyl deriv. (p. 26)	2,4-Di-nitro-phenyl deriv. (p. 27)	Notes
Aniline	184		114	163	103	156	
o-Toluidine	200		109	144	110	120	
m-Toluidine	203		66	125	114	159	
o-Ethylaniline	210		112	147			Picrate, 194 (p. 27).
2,4-Dimethylaniline (4-m-Xylidine)	212		130	192	181	156	
2,5-Dimethylaniline (2-p-Xylidine)	213	15	139	140	119 (232)	150	
2,6-Dimethylaniline (2-m-Xylidine)	215		177	168	212		
p-Ethylaniline	216		94	151	104		
o-Methoxyaniline (o-Anisidine)	218	5	85	60	127	151	
3,5-Dimethylaniline (5-m-Xylidine)	220		144	136			
2,3-Dimethylaniline (3-o-Xylidine)	221		134	189			
o-Ethoxyaniline (o-Phenetidine)	228		79	104	164	164	
2,4,6-Trimethylaniline (Mesidine)	229		216	204	167		
m-Ethoxyaniline (m-Phenetidine)	248		96	103	157		
o-Aminoacetophenone	250d	20	76	98	148		See also *Table 27.*
m-Methoxyaniline (m-Anisidine)	251		80		68	138	
p-Ethoxyaniline (p-Phenetidine)	254	2	135	173	106	118	
Methyl anthranilate	255d	25	101	100			See also *Table 19.*
p-Diethylaminoaniline	262		104	172			
Ethyl anthranilate	265d	13	61	98	112	164	See also *Table 19.*
p-Toluidine	201	45	153 (147)	158	118	136	
3,4-Dimethylaniline (4-o-Xylidine)		48	99	118	154	141	
o-Biphenylylamine (o-Phenylaniline)		49	118	102			
1-Naphthylamine (α-Naphthylamine)		50	160	161	157	190	
p-Biphenylylamine (p-Phenylaniline)		53	175	233	255		

Table 9 (cont). Amines, primary aromatic (C, H, (O) and N)

	M.p.	Acetyl deriv. (p. 26)	Benz-oyl deriv. (p. 26)	Tolu-ene-*p*-sulph-onyl deriv. (p. 26)	2,4-Di-nitro-phenyl deriv. (p. 27)	Notes
p-Dimethylaminoaniline	53	131	228			
p-Methoxyaniline (*p*-Anisidine)	57	130	154	114	141	
2-Aminopyridine	58	71	169			
m-Phenylenediamine	63	190 di 88* mono	240 di 125* mono	172	172	* More easily prepared than the di-amides.
3-Aminopyridine	64	133	119			
o-Nitroaniline	71	93	98	115 (142)		Orange.
p-Aminodiphenylamine	75	158	203		190	
2,4-Dimethyl-6-nitroaniline	76	176	185			Orange.
4-Methyl-3-nitroaniline	78	145	172	164		Yellow.
2,4-Diaminophenol	79	180 tri 220 di-*N*	253 di-*N*			
2,5-Dimethoxyaniline	83	91	85	80	186	
2,5-Diethoxyaniline	86	116	88	132		
4-Methyl-1,2-phenylenediamine (3,4-Diaminotoluene)	89	210	263	140 mono		
Ethyl *p*-aminobenzoate (Benzocaine)	90	110	148			
2-Methyl-3-nitroaniline	92	158	168			Pale yellow.
2-Methyl-6-nitroaniline	97	158	167	122		Orange.
m-Aminoacetophenone	99	128		130		See also *Table 27.*
4-Methyl-1,3-phenylenediamine (2,4-Diaminotoluene)	99	224	224	192	184	
o-Phenylenediamine	102	186	301	202		
p-Aminoacetophenone	106	167	205	203		
2-Methyl-4-nitroaniline	107	151	183			
2-Naphthylamine	112	134	162	133	179	Carcinogenic compound.
m-Nitroaniline	114	154	155	138	193	Yellow.
4-Methyl-2-nitroaniline	117	96	148	166 (146)		Red.
m-Aminophenol	122	101 di 148 mono-*N*	153 di 174 mono-*N*	157 mono		See also *Table 31.* Mono-*O*-benzoyl, 153.
2,4-Dimethyl-5-nitroaniline	123	159	200	192		

Table 9 (cont). Amines, primary aromatic (C, H, (O) and N)

	M.p.	Acetyl deriv. (p. 26)	Benz-oyl deriv. (p. 26)	Tolu-ene-*p*-sulph-onyl deriv. (p. 26)	2,4-Di-nitro-phenyl deriv. (p. 27)	Notes
p-Aminobenzophenone	124	153	152			
pp'-Diaminodiphenyl	127	317	352	243		
(Benzidine)		di	di			
		199	203			
		mono	mono			
4,4'-Diamino-3,3'dimethyl-						
diphenyl	129	314	265			Picrate, 185 (p. 27).
(*o*-Tolidine)		di	di			
		103	198			
		mono	mono			
2-Methyl-4-nitroaniline	130	202	178	174		Light yellow.
p-Phenylenediamine	140	303	338	266	177	
		di	di			
		162	128			
		mono	mono			
o-Aminobenzoic acid	144	185	182	217		See also *Table 17.*
(Anthranilic acid)						
p-Nitroaniline	147	216	199	191	186	Yellow.
2-Hydroxy-3,5-dinitroaniline	168	201	230	191		Red. See also *Table 31.*
(2-Amino-4,6-dinitrophenol,		*N*-	*N*-	*N*-		
Picramic acid)		193	220			
		O-	*O*-			
o-Aminophenol	174d	124	165	146		See also *Table 31.*
		di	*N*-	(139)		* Normal product of
		201*	185			acetylation.
		mono	*O*-			
m-Aminobenzoic acid	174	250	248			See also *Table 17.*
2,4-Dinitroaniline	180	120	202	219		Yellow; gives red colour with dil. NaOH and acetone.
p-Aminophenol	184d	150	234	168		See also *Table 31.*
		di	di	di		
		168	216	252		
		N-	*N*-	*N*-		
				142		
				O-		
p-Aminobenzoic acid	186	252	278	223		See also *Table 17.*
2,4,6-Trinitroaniline	190	230	196			Yellow.
(Picramide)						
p-Aminophenylacetic acid	200	170	198			See also *Table 17.*

Table 10. Amines, primary aromatic (C, H, (O), N and halogen or S)

	B.p.	M.p.	Acetyl deriv. (p. 26)	Benz- oyl deriv. (p. 26)	Tolu- ene-*p*- sulph- onyl deriv. (p. 26)	2,4-Di- nitro- phenyl deriv. (p. 27)	Notes
p-Fluoroaniline	188		152	185			
o-Chloroaniline	208		88	99	105	150	
2-Chloro-4-methylaniline	223		118	138	103		
m-Chloroaniline	230		73	122	138	184	
o-Aminothiophenol	234	26	135	154			
2-Bromo-4-methylaniline	240	26	118	149			
4-Chloro-2-methylaniline	241	29	140	172	145		
3-Chloro-2-methylaniline	245		159	173			
o-Bromoaniline	250 (229)	32	100	116	90	161	
m-Bromoaniline	251	18	88	135		178	
3-Bromo-4-methylaniline	254	25	113	132			
m-Iodoaniline	145/15 mm	33	119	157	128		
p-Aminothiophenol	140/16 mm	46	144 di 154 N-	180 N-			
2,5-Dichloroaniline	251	50	133	120			
2,5-Dibromoaniline		51	172				
4-Bromo-2-methylaniline	240	59	157	115			
o-Iodoaniline		60	110	139			
2,4-Dichloroaniline		63	146	117	126	116	
p-Bromoaniline		66	167	202	101	158	
p-Iodoaniline		67	184	222		176	
p-Chloroaniline		70	178	192	95	167	
2,4,6-Trichloroaniline		78	205	174			Insol. in HCl; diazotize in ethanol and H_2SO_4.
2,6-Dibromo-4-methylaniline		78	183* mono				* Diacetyl deriv., 101.
2,4-Dibromoaniline		79	146	134	134		
2,6-Dibromoaniline		83	210* mono				* Diacetyl deriv., 110 Picrate, 123 (p. 27).
4-Chloro-1,3-phenylenediamine (2,4-Diaminochlorobenzene)		88	243* di	178	215		* Monoacetyl deriv., 170.
2-Bromo-4-nitroaniline		105	129	160			Yellow.
2-Chloro-4-nitroaniline		108	139	161	164		Yellow.
4-Bromo-2-nitroaniline		111	104	137			Orange.
4-Chloro-2-nitroaniline		116	104		110		Orange.
2,4,6-Tribromoaniline		119	232 mono 127 di	198			Insol. in HCl; diazotize in ethanol and H_2SO_4.
p-Aminobenzenesulphonamide (Sulphanilamide)	165		219	284			See also *Table 33*.

Table 11. Amines, secondary

	B.p.	M.p.	Acetyl deriv. (p. 26)	Benzoyl deriv. (p. 26)	Toluene-p-sulphonyl deriv. (p. 26)	2,4-Dinitrophenyl deriv. (p. 27)	Notes
Dimethylamine	7			42	79	87	Normally in aq. solution.
Diethylamine	55			42	60	80	
Di-isopropylamine	84						Picrate, 140 (p. 27).
Pyrrolidine	89				123		Picrate, 112 (p. 27).
Piperidine	106			48	103	93	Miscible with water.
Dipropylamine	110					40	Picrate, 75 (p. 27).
(±)-2-Methylpiperidine	117			45	55		Picrate, 135 (p. 27).
(±)-3-Methylpiperidine	124						Picrate, 137 (p. 27).
Morpholine	130			75	147		
Di-isobutylamine	139		86			112	
N-Methylcyclohexyl-amine	146			85			Picrate, 170 (p. 27).
2-Ethylpiperidine	146						Picrate, 133 (p. 27).
Dibutylamine	159						Picrate, 60 (p. 27).
N-Methylaniline	194		101	63	95	167	
N-Ethylaniline	205		54	60	88	95	
N-Methyl-m-toluidine	206		66				
N-Methyl-o-toluidine	207		56	66	120	155	
N-Methyl-p-toluidine	208		83	53	60		
N-Ethyl-o-toluidine	214			72	75	114	
N-Ethyl-p-toluidine	217			40	71	120	
N-Ethyl-m-toluidine	221			72			Picrate, 132 (p. 27).
N-Propylaniline	222		48		56		
1,2,3,4-Tetrahydro-isoquinoline	233		46	129			Picrate, 200 (p. 27).
N-Butylaniline	240			56	54		
1,2,3,4-Tetrahydro-quinoline	250	20		75			Picrate, 141 (p. 27).
Dicyclohexylamine	254	20	103	153	119		
Di-(2-hydroxyethyl)amine (Diethanolamine)	270	28			99		Picrate, 110 (p. 27).
N-Methyl-1-naphthyl-amine	293			95	121	164	
Dibenzylamine	300			112	158	105	
N-Benzylaniline	306	37	58	107	148	168	
Piperazine hexahydrate		44*	134 di / 52 mono	191 di / 75 mono	173 mono		* Anhydrous, 104, and is sol. in water but only slightly in ether.
Indole		52		68			Picrate, 187.
Diphenylamine		54	101	180 (107*)	142		* Resolidifies at 135 and melts at 180.
N-Phenyl-1-naphthylamine		62	115	152			
N-Methyl-m-nitroaniline		66	95	105 (156)			
N-Phenyl-2-naphthylamine		108	93	148 (111)			
N-Methyl-p-nitroaniline		152	153	112			
Carbazole		243	69	98	137		Very weakly basic.

Table 12. Amines, tertiary

	B.p.	M.p.	Meth-iodide (p. 27)	Picrate (p. 27)	Notes
Trimethylamine	3		230	216	
Triethylamine	89		280	173	
Pyridine	116		117	167	
2-Methylpyridine (α-Picoline)	129		230	169	
2-Dimethylaminoethanol	135			96	
2,6-Dimethylpyridine (2,6-Lutidine)	143		238	161	
3-Methylpyridine (β-Picoline)	144		92	150	
4-Methylpyridine (γ-Picoline)	144		152	167	
2-Ethylpyridine	149			187	
Tripropylamine	156		208	117	
2,4-Dimethylpyridine (2,4-Lutidine)	158		113	180	
2-Diethylaminoethanol	161		249d	79	
4-Ethylpyridine	164			168	
2-Chloropyridine	166				Methyl toluene-p-sulph-onate salt, 120 (p. 28).
3,5-Dimethylpyridine (3,5-Lutidine)	170			238	
2,4,6-Trimethylpyridine (2,4,6-Collidine)	172			156	
NN-Dimethyl-o-toluidine	185		210	122	
NN-Dimethylaniline	193		218*	162	* Sublimes. p-Nitroso deriv., 87 (p. 27).
N-Ethyl-N-methylaniline	201		125	134	p-Nitroso deriv., 66 (p. 27).
NN-Dimethyl-p-toluidine	210		215*	130	* Sublimes.
Tributylamine	211		180	107	
NN-Dimethyl-m-toluidine	212		177	131	
NN-Diethylaniline	216		104	142	p-Nitroso deriv., 84 (p. 27).
NN-Diethyl-p-toluidine	229		184	110	
NN-Diethyl-m-toluidine	231			97	
Quinoline	238		72*	203	* Hydrate; anhydrous, 133.
Isoquinoline	243	24	159	222	
NN-Dipropylaniline	245		156	261	
2-Methylquinoline (Quinaldine)	247		195	191	
8-Methylquinoline	248		193	203	
6-Methylquinoline	258		219	234	
4-Methylquinoline (Lepidine)	262		174	212	
2,4-Dimethylquinoline	264		252	193	
N-Benzyl-N-methylaniline	306		164	127	p-Nitroso deriv., 44 (p. 27).
N-Benzyl-N-ethylaniline	312d		161	111	p-Nitroso deriv., 62.

Table 12 (cont). Amines, tertiary

	B.p.	M.p.	Meth-iodide (p. 27)	Picrate (p. 27)	Notes
p-Diethylaminobenzaldehyde		41			Yellow; see *Table 5.*
NN-Dimethyl-p-bromoaniline		55	185		
2,6-Dimethylquinoline		60	244d	191	
NN-Dimethyl-m-nitroaniline		60	205	119	Red.
NN-Dibenzylaniline		71	135	131	p-Nitroso deriv., 91 (p. 27).
p-Dimethylaminobenzaldehyde		74			See *Table 5.*
NN-Dimethyl-p-aminophenol		75			See *Table 31.*
8-Hydroxyquinoline		75	143	204	See also *Table 31.*
NN-Dimethyl-m-aminophenol		85	182	162	p-Nitroso deriv., 169; see also *Table 31.*
Tribenzylamine		92	184	190	
Acridine		110	224	208	
2,3-Dimethyl-1-phenylpyrazol-5-one (Phenazone, antipyrine)		113		181	p-Nitroso deriv., 200. Absorbs bromine; gives orange colour with aq. FeCl$_3$.
Triphenylamine		127			Not basic; nitration in acetic acid gives trinitro deriv., 280.
pp'-Bis-(dimethylamino)benzophenone (Michler's ketone)		174	105	156	Oxime, 233 (p. 24). See also *Table 27.*
Hexamethylene tetramine		280*	190	179	* Sublimes.

65

Table 13. Amino-acids

	Decomposition temp.	Benzoyl deriv. (p. 28)	3,5-Dinitrobenzoyl deriv. (p. 28)	Toluene-p-sulphonyl deriv. (p. 28)	Notes
N-Phenylglycine	126	63			Acetyl deriv., 194 (p. 26).
(+)- or (−)-Ornithine	140*	188 di 240 mono			* Often a syrup. Picrate, 204 (p. 26).
o-Aminobenzoic acid (Anthranilic acid)	144	182	278		Acetyl deriv., 185 (p. 26); see also *Tables 9* and *17*.
m-Aminobenzoic acid	174	248			Acetyl deriv., 250 (p. 26); see also *Tables 9* and *17*.
p-Aminobenzoic acid	186	278			Acetyl deriv., 252 (p. 26); see also *Tables 9* and *17*.
3-Aminopropionic acid (β-Alanine)	196		202	117	
(±)-Glutamic acid	199	157		117	Acetyl deriv., 185 (p. 26).
p-Aminophenylacetic acid	200	205			Acetyl deriv., 170 (p. 26).
(±)-Proline	203*		217		* Monohydrate, 190. Picrate, 135.
(+)- or (−)-Arginine	207	298 mono 235 di	150		Picrate, 206 (p. 27).
(+)- or (−)-Glutamic acid	211	138	217	131	
Sarcosine	212	103	154	102	
(+)- or (−)-Proline	222	156		133	Picrate, 154 (p. 27).
(+)- or (−)-Lysine	224	149	169		Picrate, 266 (p. 27).
(+)- or (−)-Asparagine	226	189	196	175	
(±)-Serine	228	171 mono 124 di	95	213	
Glycine	232	187	179	147	Acetyl deriv., 206 (p. 26).
(±)-Threonine	235	174* di 176* mono			* Mixed m.p. 145.
(±)-Arginine	238	230*			* Hydrate, 176. Picrate, 201 (mono), 196 (di) (p. 27).
(+)- or (−)-Cystine	260	181	180	201	
(+)- or (−)-Aspartic acid	271	185		140	
(±)-Methionine	281 (272)	151		105	Acetyl deriv., 114 (p. 26).
(±)-Phenylalanine	273	188	93	134	

Table 13 (cont). Amino-acids

	Decomposition temp.	Benzoyl deriv. (p. 28)	3,5-Dinitrobenzoyl deriv. (p. 28)	Toluene-*p*-sulphonyl deriv. (p. 28)	Notes
(±)-Tryptophan	275	188	240	176	
(+)- or (−)-Histidine	277	249	189	203	
(±)-2-Amino-2-methylprop-ionic acid	280*	198			* Sublimes.
(±)-Aspartic acid	280	165			
(+)- or (−)-Methionine	283	150	95		Acetyl deriv., 98 (p. 26).
(+)- or (−)-Tryptophan	289	176	233d	176	
(±)-Isoleucine	292	118		140	
(±)-Alanine	295	166	177	139	
(+)- or (−)-Alanine	297	151		133	
(±)-Valine	298	132	158	110	
(±)-2-Aminobutanoic acid	307	147	194		
(+)- or (−)-Valine	315	127	158	147	
(±)-Tyrosine	318	197 mono-*N*	254	224	
(+)- or (−)-Phenylalanine	320	146	93	164	
(±)-Norleucine	327			124	
(±)-Leucine	332	141	187		Acetyl deriv., 157.
(+)- or (−)-Leucine	337	107*	187	124	* Hydrate, 60.
(+)- or (−)-Tyrosine	344	166 mono-*N* 211 di		188 mono-*N* 119 di	Acetyl deriv., 172
(±)-Ornithine		267d mono 188 di		188 mono	
(±)-Lysine		249 mono 145 di			Picrate, 225d (mono).

Table 14. Azo, Azoxy, Nitroso and Hydrazine Compounds

	B.p.	Notes
NN-Dimethylhydrazine	63	Picrate, 146 (p. 27).
NN'-Dimethylhydrazine	81	Picrate, 148 (p. 27).
N-Methyl-N-phenylhydrazine	227	Benzoyl deriv., 153; acetyl deriv., 92 (p. 26); with benzaldehyde → hydrazone, 106.
N-Ethyl-N-phenylhydrazine	237	Hydrochloride, 146.
N-Ethyl-N'-phenylhydrazine	238	Benzoyl deriv., 100 (p. 26).
Phenylhydrazine (m.p. 19)	243	Benzoyl deriv., 168; with benzaldehyde → hydrazone, 158.

	M.p.	
NN-Diphenylhydrazine	34	Benzoyl deriv., 192; acetyl deriv., 184; with benzaldehyde → hydrazone, 122.
Azoxybenzene	36	Light yellow. Warming with conc. H_2SO_4 → p-hydroxyazobenzene, 152.
p-Tolylhydrazine	66	Benzoyl deriv., 146; acetyl deriv., 121; with benzaldehyde → hydrazone, 125.
Nitrosobenzene	68	Colourless solid but turns green on melting. With Sn + HCl → aniline (p. 41).
Azobenzene	68	Orange-red colour. Zn dust + ethanolic NaOH → hydrazobenzene, 131.
p-Azoxytoluene	75	Light yellow. Zn dust + ethanolic NaOH → p-azotoluene, 144.
NN-Diethyl-p-nitrosoaniline	84	Dark green, forms yellow hydrochloride; $KMnO_4$ → NN-diethyl-p-nitroaniline, 77.
NN-Dimethyl-p-nitrosoaniline	85	Dark green, forms yellow hydrochloride; $KMnO_4$ → NN-dimethyl-p-nitroaniline, 163.
p-Nitrosophenol	125d	Colourless. Acetic anhydride at 100° → acetyl deriv., 107 (yellow).
1,2-Diphenylhydrazine (Hydrazobenzene)	127	Dibenzoyl deriv., 162 (p. 26).
Benzeneazo-2-naphthol	133	Orange colour. See Table 31.
p-Nitrosodiphenylamine	145	Green. Zn + acetic acid → p-amino deriv., 66.
1,2-Di-(o-hydroxyphenyl)hydrazine (2,2'-Hydrazophenol)	148	Yellow. Benzoyl deriv., 186; see also Table 31.
p-Hydroxyazobenzene	152	Yellow. Acetyl deriv., 84; benzoyl deriv., 138 (p. 26).

Table 15. Carbohydrates

	Approximate decomposition temperature	$[\alpha]_D$ (in water) Initial	$[\alpha]_D$ (in water) Final	p-N-Glycosyl-aminobenzoic acid (p. 30)	Acetate* (p. 29)	Osazone (p. 31)	Notes
Melibiose, monohydrate	85	+111	+129		α147 β177	178	Disaccharide
D-Ribose	90	−21	−21	156		164	
D-Glucose, monohydrate	90	+112	+52	134	α112 β132	205d	Pentabenzoate, 179. (p. 22)
Maltose, monohydrate	101	+111	+130		α125 β160	206	Disaccharide.
D-Fructose	104	−132	−92		α 70 β109	205d	Pentabenzoate, 79. (p. 22)
L-Rhamnose, monohydrate	105	−8·6	+8·2	170	99	182	
L-Rhamnose, anhydrous	123	−8·6	+8·2	170	99	182	
D-Mannose	132	+29	+14	182	α 74 β115	205d	
D-Xylose	144	+93	+18	181	α 59 β126	160	
D-Glucose, anhydrous	146	+112	+52	134	α112 β132	205d	
D- or L-Arabinose	160	−175 D- +190 L-	−105 D- +104 L-	192	α 94 β 86	166	
L-Sorbose	161	−43	−43		97	162	
Maltose, anhydrous	165	+111	+130		α125 β160	206	Disaccharide.
D-Galactose	170	+150	+80	160	α 95 β142	196	
Sucrose	185	+66	+66		70	205	Non-reducing disaccharide.
Lactose, monohydrate	203	+90	+55		α152 β 90	200d	Disaccharide.
Cellobiose	225	+14	+35		α229 β192	200	Disaccharide.

* The preparation of the β-form is described on p. 29 but the melting points of both anomers are given because a small amount of the α-form is sometimes formed.

Table 16. Carboxylic acids (C, H and O), their acyl chlorides, anhydrides and nitriles

	B.p.	M.p.	Chloride B.p.	Anhydride B.p.	Nitrile B.p.	Amide (p. 31) M.p.	Anilide (p. 31) M.p.	p-Toluidide (p. 31) M.p.	p-Bromophenacyl ester (p. 31) M.p.	p-Phenylphenacyl ester (p. 31) M.p.	Notes
Formic	100	8	—	—	26	3	50*	53*	140†	74	* By heating the acid with the amine. † ArCO·CH$_2$·OH is often isolated; ester melts at 99.
Acetic	118	16	52	140	82	82	114	153 (147)	85	111	
Propionic	140	13	80	168	97	81	106	126	61	102	
Propenoic (Acrylic)	140		75		78	84	104	141			Unsaturated; polymerizes readily.
Propynoic (Propiolic)	144d	18				61	87				Unsaturated.
2-Methylpropionic (Isobutyric)	155		92	182	108	128	105	108	77	89	
2-Methylpropenoic (Methacrylic)	161	16	95		90	102	87				Unsaturated
Butanoic (Butyric)	163		101	198	118	116	96	75	63	97	
2,2-Dimethylpropionic (Pivalic)	164	35	105	190	106	155	132	119	76	114	
2-Oxopropionic (Pyruvic)	165	13	98			124	104	109			See also Table 26.
But-3-enoic (Vinylacetic)	169 (163)					73	58		60		Unsaturated.
cis-But-2-enoic (Isocrotonic)	169					102	101	132	81		Unsaturated.

70

Acid											Notes
(±)-2-Methylbutanoic	177		115		125	112	110	93	55	71	
3-Methylbutanoic (Isovaleric)	177		115	215	129	135	110	107	68	78	
Pentanoic (Valeric)	186		126	218	140	106	63	74	75	64	
2-Ethylbutanoic	195		139	229	145	112	127	116	77	77	
4-Methylpentanoic (Isocaproic)	199				155	120	112	63		70	
Hexanoic (Caproic)	205		153	254	163	100	94	74	72	68	
Heptanoic	223		193	258	184	96	65	81	72	62	
2-Ethylhexanoic	227		88/20 mm	150/8 mm	75/9 mm	102	57			53	
Octanoic (Caprylic)	237	16	195	281	205	106	57	70	66	67	See also Table 26.
4-Oxopentanoic (Laevulinic, laevulic)	245d	33				107	102	109	84		
Nonanoic	254	12	215		224	101	57	84	69	71	
(±)-2-Phenylpropionic	265		97/12 mm		230	95	67	68	66		
Decanoic (Capric)	269	31	114/15 mm	24*	245	108	70	78	66	80	* M.p.
Undec-10-enoic (Undecylenic)	275	24		37*	248	87	67	68		61	Unsaturated.
Undecanoic	280	28		22*		103	71	80	68		* M.p.
cis-Octadec-9-enoic (Oleic)	286d	16			330d	76	41	42	46		* M.p. Unsaturated.
(±)-Lactic	122/15 mm	18			182	78	58	107	113	145	Gives positive iodoform reaction.
Dodecanoic (Lauric)		44	145/18 mm	42*	280	100	78	87	76	84	* M.p.
3-Phenylpropionic (Hydrocinnamic)		48	225d		261	105 (82) 93	96	135	104	95	
trans-Octadec-9-enoic (Elaidic; trans-Oleic)		51		51*					65	73	* M.p. Unsaturated.
4-Phenylbutanoic (4-Phenylbutyric)		52	119/9 mm			84			58	90	

Table 16 (cont). Carboxylic acids (C, H and O), their acyl chlorides, anhydrides and nitriles

	M.p.	Chloride B.p.	Anhydride B.p.	Nitrile B.p.	Amide (p. 31) M.p.	Anilide (p. 31) M.p.	p-Toluidide (p. 31) M.p.	p-Bromophenacyl ester (p. 31) M.p.	p-Phenylphenacyl ester (p. 31) M.p.	Notes
Tetradecanoic (Myristic)	54	174/16 mm	54*	19*	102	84	93	81	90	* M.p.
Hexadecanoic (Palmitic)	62	12*	63*	31*	106	89	98	84	94	* M.p.
3-Methylbut-2-enoic (ββ-Dimethylacrylic)	68	145		140	107			104	145	Unsaturated.
Octadecanoic (Stearic)	70	22*	70*	43*	108	94	102	90	97	* M.p.
But-2-enoic (Crotonic)	72	126	246	118	158	118	132	96		Unsaturated.
Phenylacetic	76	183	72*	232	157	118	136	89	63d	* M.p.
2-Hydroxy-2-methylpropionic (α-Hydroxybutyric)	79				98	136	133			
Hydroxyacetic (Glycollic)	80		128*	183	120	97	143	138		Is often a syrup. * M.p.
Methylmaleic (Citraconic)	92d	95/18 mm	214		185	176 di / 153 mono	170 mono		109	
Glutaric	98	218	56*	286	175	224	218	137	152	* M.p.
Phenoxyacetic	99	225	67*	239	101	99		149	146	* M.p.
Citric, monohydrate	100				210	199	189	148		Gives positive iodoform test. Heat at 130° → anhyd. acid, 153.

72

Compound									
Oxalic, dihydrate	100	64		419 di / 219 mono	246 di / 148 mono	268 di / 169 mono	242	166	Heat → anhydrous acid, 189.
(−)-Hydroxysuccinic (Malic)	100			156	197 mono	207 mono	179	106	
o-Methoxybenzoic (o-Anisic)	100	254	24*	129	131		113	131	* M.p.
Heptanedioic (Pimelic)	104			175	156 di / 108 mono	206	137	146d	
o-Toluic	105	212; 166/18 mm	39*	142	125	144	57	95	* M.p.
Nonanedioic (Azelaic)	106		205	175 di / 94 mono	187 di / 107 mono	200	131	141	
m-Toluic	111	218	71*	96	126	118	108	137	* M.p.
3-Benzoylpropionic	116		212	125	150		98		
(±)-3-Hydroxy-2-phenylpropionic ((±)-Tropic)	117			169 mono					Acetate, 88 (p. 22).
(±)-Mandelic	118	197; 70†	21*	133	151	172	113	167	* M.p.
Benzoic	122		191	128	163	158	119		* M.p.
o-Benzoylbenzoic	128*		42*	165	195			168	* Monohydrate, 91.
Maleic	130 (139)	60*	56* / 31*	266 di / 181 mono	187	142	168		† M.p. * M.p.
Malonic	133d		219	170 di / 50 mono	225	252		175	
Decanedioic (Sebacic)	133	182/16 mm		210 di / 170 mono	200 di / 122 mono	201 di / 122 mono	147	140	

Table 16 (cont). Carboxylic acids (C, H and O), their acyl chlorides, anhydrides and nitriles

	M.p.	Chloride B.p.	Anhydride B.p.	Nitrile B.p.	Amide (p. 31) M.p.	Anilide (p. 31) M.p.	p-Toluidide (p. 31) M.p.	p-Bromophenacyl ester (p. 31) M.p.	p-Phenylphenacyl ester (p. 31) M.p.	Notes
Cinnamic	133	35*	136*	255 / 20*	147	151	168	146	183	Unsaturated.
Furoic	133	173	73*	146	142	124	108	139	91	* M.p.
1-Naphthylacetic	133	188/23 mm		183	181	155		112		* M.p.
Hexa-2,4-dienoic (Sorbic)	134	78/15 mm			168	153		129	141	Unsaturated.
Acetylsalicylic (Aspirin)	135		43*		138	136				* M.p.
3-Phenylpropynoic (Phenylpropiolic)	137	116/17 mm		41*	100	126	142			Unsaturated. * M.p.
meso-Tartaric	140			131*	190	193 mono				* M.p.
Octanedioic (Suberic)	142	162/15 mm	65*		216 di 125 mono	187 di 128 mono	219	144	151	* M.p.
3,4,5-Trimethoxybenzoic	144				184	154		129		
Diphenylacetic	148	56*	98*	72*	168	180	172	112	111	* M.p.
Benzilic (Diphenylglycollic)	150	193/27 mm			154	175	189	152	122	
Citric, anhydrous	153				210*	199*	189*	148	146	Gives positive iodoform test. * From amine and ester on prolonged heating.

Name	M.p.									Notes
Hexane-1,6-dioic (Adipic)	153	130/18 mm	295		220 di / 126 mono / 177 mono	238 di / 151 mono	241	155	148	See also *Table 30.*
5-Methylsalicylic (p-Cresotic)	153							142		See also *Table 30.*
Salicylic	158	20*	98*		139	135	156	140	148	See also *Table 30.* * M.p. * M.p.
1-Naphthoic (α-Naphthoic)	161	145*			202	163			126	See also *Table 30.*
3-Methylsalicylic (o-Cresotic)	163		35*		112			135		See also *Table 30.*
Methylenesuccinic (Itaconic)	165				192	190*	190*	117		* By heating acid with an excess of amine. Unsaturated.
(±)-Phenylsuccinic	167				211 / 195 di / 171 mono	222 / 264 di / 180 mono				
(+)-Tartaric	169							216	204d	
4-Methylsalicylic (m-Cresotic)	177									See also *Table 30.*
Butynedioic (Acetylenedicarboxylic)	179				294d					Unsaturated.
p-Toluic	180	95*	218	214	160 / 164	144 / 154		153 / 124	165	* M.p.
3,4-Dimethoxybenzoic (Veratric)	181	99*	61*	22*	163	169	186		160	* M.p.
p-Methoxybenzoic (p-Anisic)	184	141*			228	232		152		* M.p.
o-Carboxyphenylacetic (Homophthalic)	185	133*	66*	43*						* M.p.
2-Naphthoic (β-Naphthoic)	185				192	171	191	211	183	* M.p.
Succinic	185	119*	54*	20*	260 di / 157 mono	230 di / 148 mono	255 di / 179 mono	211	208	* M.p.

Table 16 (cont). Carboxylic acids (C, H and O), their acyl chlorides, anhydrides and nitriles

	M.p.	Chloride B.p.	Anhydride B.p.	Nitrile B.p.	Amide (p. 31) M.p.	Anilide (p. 31) M.p.	p-Toluidide (p. 31) M.p.	p-Bromophenacyl ester (p. 31) M.p.	p-Phenylphenacyl ester (p. 31) M.p.	Notes
(+)-Camphoric	187				192 di 177 mono	226 di 204 mono				
Oxalic, anhydrous	188	64			419 di 219 mono	246 di 148 mono	268 di 169 mono	242	166	
1-Hydroxy-2-naphthoic	195	85*			202	154				* M.p. Acetate, 158.
m-Hydroxybenzoic	200			82*	170	156	163	176 (168)		* M.p. See also Table 30.
3,4-Dihydroxybenzoic (Protocatechuic)	200d				212	166				
2,5-Dihydroxybenzoic (Gentisic)	200	98*								* M.p. See Table 30.
Phthalic	200d	275	131*	141*	220† mono	254 di 170 mono	201 di 155 mono	153	167	Loses water near m.p. to form anhydride. * M.p. † Loses NH₃ near m.p. to form imide, 233.

Name									
(±)-Tartaric	205								
p-Hydroxybenzoic	213		113*	226 / 162†	235 / 196†	204†	191	240	* M.p. † Difficult to prepare. See also Table 30.
2,4-Dihydroxybenzoic (β-Resorcylic)	213d			222	126		225		
Mucic	214d			220 di / 192 mono				149d	Tetra-acetate, 266 (p. 22).
3-Hydroxy-2-naphthoic	222	95*	188*	217	243	221	134		* M.p. Acetate, 184.
3,4,5-Trihydroxybenzoic (Gallic)	240d			245* (189)	207*			198d	* Difficult to prepare. Triacetate, 172; tribenzoate, 192.
Fumaric	287*	160	96†	267	314		256d		* In a sealed tube; sublimes at 200.
Terephthalic	>300*	83†	222†	>350d	334		225		† M.p. * Sublimes.
Isophthalic	345	41*	162*	280	250		179	280	† M.p. * M.p.

Table 17. Carboxylic acids (C, H, O and halogen, N or S)

	B.p.	M.p.	Amide (p. 31)	Anilide (p. 31)	p-Toluidide (p. 31)	p-Bromophenacyl ester (p. 31)	p-Phenylphenacyl ester (p. 31)	Notes
Trifluoroacetic	72		74	91				Acid chloride b.p. −27.
Thioacetic	93		108	76	130			Pale yellow; unpleasant odour.
Fluoroacetic	167	31	108					
2-Chloropropionic	186		80	92	124			
Dichloroacetic	194	5	98	125	153	99		
2-Bromopropionic	203	25	123	99	125			
2-Bromobutanoic (2-Bromobutyric)	217d		112	98	92			
Thioglycollic	123/29 mm		52	111	125			
3-Chloropropionic		40	101					
2-Bromo-3-methyl-butanoic (α-Bromoisovaleric)		44	133	116	124			
2-Bromo-2-methyl-propionic (α-Bromoisobutyric)		48	148	83	93			
Bromoacetic	208	50	91	131	91			
Trichloroacetic	196	57	141	94	113			
Chloroacetic		63	120	137	162	105	116	
Cyanoacetic		66	123	198				
Iodoacetic		83	95	143				
3-Nitrosalicylic, hydrate		125	145					
Pyridine-2-carboxylic (2-Picolinic)		138	107	76	104			
m-Nitrobenzoic		140	142	154	162	137	153	
o-Nitrophenylacetic		141	161					
4-Chloro-3-nitrobenzoic		142	172	131				
o-Chlorobenzoic		142	142	118	131	106	123	
3-Nitrosalicylic, anhyd.		144	145					
o-Aminobenzoic (Anthranilic)		144	109	131	151	172*		* Two moles of reagent are required to form this complex derivative. See also *Tables 9* and *13*.
o-Nitrobenzoic		146	175	155	203	107	140	
o-Bromobenzoic		150	155	141		102	98	
p-Nitrophenylacetic		152	198	198	210	207		

Table 17 (cont). Carboxylic acids (C, H, O and halogen, N or S)

	B.p.	M.p.	Amide (p. 31)	Anilide (p. 31)	p-Toluidide (p. 31)	p-Bromophenacyl ester (p. 31)	p-Phenylphenacyl ester (p. 31)	Notes
m-Bromobenzoic		155	155	146		126	155	
m-Chlorobenzoic		158	134	124		117	154	
2,4-Dichlorobenzoic		160	194					Conc. HNO_3 + H_2SO_4 → 3,5-dinitro deriv., 212.
o-Iodobenzoic		162	184	142		110	143	
5-Bromosalicylic		165	232	222				Acetate, 168 (p. 41).
4-Nitrophthalic		165	200d	192	172		120	
Thiosalicylic		165						Acetate, 125 (p. 41).
m-Aminobenzoic		174	111	140		190*		See also *Tables 9* and *13*. * Two moles of reagent are required to form this complex derivative.
4-Chloro-3-nitrobenzoic		181	156	131				
p-Fluorobenzoic		182	154					
2,4-Dinitrobenzoic		183	203			158		
p-Aminobenzoic		186	114			200*		See also *Tables 9* and *13*. * See under m-amino-benzoic acid.
m-Iodobenzoic		187	186			128	147	
N-Benzoylglycine (Hippuric)		187	183	208		151	163	
3,5-Dinitrobenzoic		204	183	235	280	159	154	Acyl chloride, 70.
3-Nitrophthalic		218	201d	234	224	166	149	Anhydride, 162.
5-Nitrosalicylic		229	225	224				
Pyridine-3-carboxylic (Nicotinic)		237	122	132*	150			* From benzene; from water, 85.
o-Nitrocinnamic		239	185			142	146	Unsaturated.
p-Nitrobenzoic		240	201	211	203	134	182	Acyl chloride, 75.
p-Chlorobenzoic		241	179	194		126	160	
p-Bromobenzoic		251	189	197		134	160	
p-Iodobenzoic		265	217	210		146	171	
p-Nitrocinnamic		285	217			191	192	Unsaturated.
Pyridine-4-carboxylic (Isonicotinic)		324*	155					* Sublimes.

Table 18. Enols

	B.p.	Semi-carb-azone (p. 32)	2,4-Di-nitro-phenyl-hydraz-one (p. 32)	Colour with aq. FeCl$_3$	Notes
Pentan-2,4-dione (Acetylacetone)	139	107*	122* 209 di	Red	* Pyrazole deriv. See also *Table 26.*
Methyl 3-oxobutanoate (Methyl acetoacetate)	170	152	119	Red	See also *Table 26.*
Ethyl 3-oxobutanoate (Ethyl acetoacetate)	181d	133	96	Red	See also *Table 26.*
Ethyl acetonedicarboxylate	250	94	86	Red	See also *Table 26.*
	M.p.				
Benzoylacetone	61		151	Red	See also *Table 26.*
1,3,5-Trihydroxybenzene (Phloroglucinol)	217			Violet*	* Transient colour. Picrate, 101. See also *Table 30.*

Table 19. Esters, carboxylic

	B.p.	Notes
Methyl formate	32	
Ethyl formate	54	
Methyl acetate	57	
Isopropyl formate	68	
Vinyl acetate	72	Unsaturated; polymerizes readily.
Methyl chloroformate	73	Very reactive chlorine atom.
Ethyl acetate	77	
Methyl propionate	79	
Propyl formate	81	
Allyl formate	83	Unsaturated.
Methyl acrylate	85	Unsaturated; polymerizes readily.
Isopropyl acetate	91	
Methyl 2-methylpropionate (Methyl isobutyrate)	92	
Ethyl chloroformate	93	Very reactive chlorine atom.
s-Butyl formate	97	
Isobutyl formate	98	
Ethyl propionate	98	
t-Butyl acetate	98	
Methyl methacrylate	99	Unsaturated; polymerizes readily.
Propyl acetate	101	
Ethyl acrylate	101	Unsaturated; polymerizes readily

Table 19 (cont). Esters, carboxylic

	B.p.	Notes
Methyl butanoate	102	
Allyl acetate	103	Unsaturated.
Isopropyl chloroformate	105	Very reactive chlorine atom.
Butyl formate	107	
Ethyl 2-methylpropionate (Ethyl isobutyrate)	110	
s-Butyl acetate	111	
Isopropyl propionate	111	
Isobutyl acetate	116	
Methyl 3-methylbutanoate (Methyl isovalerate)	116	
Ethyl butanoate	120	
Propyl propionate	122	
Isopentyl formate	123	
Butyl acetate	125	
Isopropyl butanoate	128	
Methyl pentanoate (Methyl valerate)	130	
Isobutyl chloroformate	130	Very reactive chlorine atom.
Methyl chloroacetate	130	
Pentyl formate	130	
Ethyl 3-methylbutanoate (Ethyl isovalerate)	134	
Methyl pyruvate	136	2,4-Dinitrophenylhydrazone, 187 (p. 23). See also *Table 26*.
Isobutyl propionate	137	
Ethyl but-2-enoate (Ethyl crotonate)	138	Unsaturated.
Isopentyl acetate	142	
Propyl butanoate	143	
Butyl propionate	144	
Methyl lactate	144	
Methyl bromoacetate	144d	
Ethyl pentanoate	145	
Ethyl chloroacetate	145	
Pentyl acetate	146	
Ethyl 2-chloropropionate	146	
Benzyl chloroacetate	147	
Isobutyl 2-methylpropionate	147	
Methyl hexanoate	150	
Ethyl lactate	152	
Ethyl pyruvate	155	Semicarbazone, 206d; 2,4-dinitrophenylhydrazone, 155 (p. 23).
Isobutyl butanoate	157	
Ethyl bromoacetate	159	
Ethyl glycollate	160	
Cyclohexyl formate	162	

Table 19 (cont). Esters, carboxylic

	B.p.	Notes
Ethyl 2-bromopropionate	162	
Butyl butanoate	165	
Ethyl pentanoate	166	
Ethyl trichloroacetate	167	
Propyl pentanoate	167	
Isopropyl lactate	168	
Pentyl propionate	169	
Methyl 3-oxobutanoate (Methyl acetoacetate)	170	Red colour with aq. $FeCl_3$; semicarbazone, 152 (p. 32); see also *Table 26*.
Methyl heptanoate	173	
Cyclohexyl acetate	175	
Butyl chloroacetate	175	
Isopentyl butanoate	178	
Ethyl 3-bromopropionate	179	
Methyl 2-furoate	181	
Methyl malonate	181	
Ethyl 3-oxobutanoate (Ethyl acetoacetate)	181	Red colour with aq. $FeCl_3$; semicarbazone, 129 (p. 32); see also *Table 26*.
Pentyl butanoate	185	
Propyl butanoate	186	
Ethyl oxalate	186	
Butyl lactate	188	
Ethyl heptanoate	189	
Isopropyl oxalate	190	
Isopentyl 3-methylbutanoate (Isopentyl isovalerate)	190	
Heptyl acetate	192	
Methyl octanoate	193	
Tetrahydrofurfuryl acetate	194	
Methyl succinate	195	
Methyl 4-oxopentanoate (Methyl laevulinate)	196	See also *Table 26*.
Phenyl acetate	196	
Methyl benzoate	198	
Ethyl malonate	199	
Methyl cyanoacetate	200	
Benzyl formate	203	
Pentyl pentanoate	204	
γ-Butyrolactone	204	
Ethyl 4-oxopentanoate (Ethyl laevulinate)	205	Semicarbazone, 135.
Methyl maleate	205	Unsaturated.
Ethyl octanoate	206	
γ-Pentanolactone	207	
o-Cresyl acetate	208	
Phenyl propionate	211	
Ethyl benzoate	212	

Table 19 (cont). Esters, carboxylic

	M.p.	Notes
m-Cresyl acetate	212	
p-Cresyl acetate	212	
Propyl oxalate	213	
Methyl *o*-toluate	213	
Benzyl acetate	214	
Methyl *m*-toluate	215	
Ethyl fumarate	216	Unsaturated.
Ethyl succinate	216	
Methyl *p*-toluate (m.p. 33)	217	
Isopropyl benzoate	218	
Methyl phenylacetate	220	
Methyl salicylate	224	See *Table 30.*
Ethyl maleate	225	Unsaturated.
Phenyl butanoate	227	
(—)-Menthyl acetate	227	
Ethyl phenylacetate	227	
Allyl benzoate	230	Unsaturated.
Ethyl salicylate	233	See *Table 30.*
Isopropyl salicylate	238	See *Table 30.*
Benzyl butanoate	238	
Propyl salicylate	239	See *Table 30.*
Isobutyl benzoate	241	
Butyl oxalate	243	
Methyl *o*-bromobenzoate	244	
Ethyl adipate	245	
Propyl succinate	246	
Methyl undecyl-10-enoate	248	Unsaturated.
Butyl benzoate	249	
Ethyl acetonedicarboxylate	250	Red colour with aq. $FeCl_3$; semicarbazone, 94; see also *Table 26.*
Ethyl mandelate (m.p. 37)	254	
Methyl anthranilate	255d	See *Table 9.*
Pentyl benzoate	255	
Ethyl benzoylformate	257	
Glyceryl triacetate	258	
Methyl dodecanoate (Methyl laurate)	258	
Butyl salicylate	259	See *Table 30.*
Glyceryl monoacetate	260	
Isobutyl salicylate	261	See *Table 30.*
Methyl cinnamate (m.p. 33)	263	Unsaturated.
Ethyl anthranilate	265d	See *Table 9.*
Ethyl *p*-methoxybenzoate (Ethyl *p*-anisate)	269	
Ethyl cinnamate (m.p. 12)	271	Unsaturated.
Ethyl tartarate (m.p. 17)	280	

Table 19 (cont). Esters, carboxylic

	B.p.	Notes
Methyl phthalate	282	
Methyl decanedioate (m.p. 38) (Methyl sebacate)	290	
Ethyl citrate	294	
Ethyl phthalate	298	
Isopropyl phthalate	302	
Ethyl isophthalate	302	
Ethyl decanedioate (Ethyl sebacate)	306	
Ethyl 2-naphthoate (m.p. 32)	308	
Ethyl 1-naphthoate	309	
Benzyl benzoate	323	
Butyl phthalate	338	
Ethyl octadecanoate (m.p. 35) (Ethyl stearate)	200/10 mm	
Methyl octadecanoate (m.p. 38) (Methyl stearate)	214/15 mm	

	M.p.	
Glyceryl 1,3-diacetate	40	
Benzyl phthalate	42	
Phenyl salicylate	42	See *Table 30.*
Cyclohexyl oxalate	42	
Ethyl terephthalate	43	
Methyl *p*-chlorobenzoate	44	
Benzyl succinate	45	
Methyl *p*-methoxybenzoate (Methyl *p*-anisate)	45	
Ethyl 3-nitrophthalate	46	
Ethyl *m*-nitrobenzoate	47	
(+)-Methyl tartarate	48	
1-Naphthyl acetate	49	
Methyl oxalate	52	
Ethyl *p*-nitrobenzoate	56	
Methyl mandelate	58	Benzoate, 76 (p. 22).
Methyl 4-nitrophthalate	66	
Methyl isophthalate	67	
Phenyl benzoate	68	
Methyl *m*-hydroxybenzoate	70	Violet colour with aq. $FeCl_3$.
2-Naphthyl acetate	70	
Glyceryl tristearate	71	
Phenyl cinnamate	72	Unsaturated.
Ethyl *m*-hydroxybenzoate	72	Violet colour with aq. $FeCl_3$. Benzoate, 58 (p. 22).
Methyl *m*-nitrobenzoate	78	
Methyl citrate	79	
Benzyl oxalate	80	
Ethyl *p*-aminobenzoate	90	See *Table 9.*

Table 21 (*cont*). Ethers

	B.p.	M.p.	Alkyl 3,5-di-nitrobenzoate (p. 34)	Picric acid complex (p. 35)	Sulphon-amide (p. 35) position	Sulphon-amide (p. 35) M.p.	Nitro deriv. (p. 35) position	Nitro deriv. (p. 35) M.p.	Nitro deriv. (p. 35) method	Notes
o-Chloroethoxy-benzene (o-Chlorophene-tole)	208	17			4	133	4	82	iii	
Butyl phenyl ether	210			112	4	104				
p-Chloroethoxy-benzene (p-Chlorophene-tole)	212	21			2	134	2,6	54	iii	
m-Dimethoxy-benzene	217			57	4	166	2,4,6	124	ii	
o-Bromoethoxy-benzene (o-Bromophene-tole)	218				4	134	4	98	iii	
o-Bromomethoxy-benzene (o-Bromoanisole)	218				4	140	4	106	iii	
p-Bromomethoxy-benzene (p-Bromoanisole)	223	12			2	148	2	88	iii	
Dihexyl ether	228									
p-Bromoethoxy-benzene (p-Bromophene-tole)	229	12	61		2	144	2	47	iii	
4-Allyl-1,2-methyl-enedioxybenzene (Safrole)	232	11		104			1,3,5	51	iii	Unsaturated; Br₂ in ether → penta-bromo deriv., 169 (p. 35).
m-Diethoxybenzene	235	12		109	4	184				Br₂ in acetic acid → tribromo deriv., 69 (p. 35).
p-Prop-1-enylmeth-oxybenzene (Anethole)	235	21		70						Unsaturated; tri-bromo deriv., 108 (p. 35); CrO₃-acetic acid → p-methoxybenzoic acid, 184 (p. 23).

87

Table 21 (cont). Ethers

	B.p.	M.p.	Alkyl 3,5-di-nitrobenzoate (p. 34)	Picric acid complex (p. 35)	Sulphonamide (p. 35) position	M.p.	Nitro deriv. (p. 35) position	M.p.	method	Notes
Diphenyl ether	259	28		110	4,4'	159	4,4'	144	iii	
1-Methoxynaph-thalene	271			130	4	156	2,4,5	128	iii	
1-Ethoxynaph-thalene	280	5		119	4	164	2,4,5	149	iii	
2-Ethoxynaph-thalene	282	37		101	8	162	1,6,8	186	iii	
Dibenzyl ether	295d	3	112	78						Dibromo deriv., 107 (p. 35).
o-Diethoxybenzene	217	43		71	4	162	4	73	v	
							3,4,6	122	ii	
1,2,3-Trimethoxy-benzene	241	47		81	2,3,4	123	5	106	ii	
p-Dimethoxy-benzene	212	56		48	2	148	2	72	i	
p-Diethoxybenzene		72			2	154	2	49	i	
2-Methoxynaphthalene		72		117	8	151	1,6,8	215d	iii	
p-Dibenzyloxybenzene		127					2	83	i	

Table 22. Halides, alkyl mono-

	Chloride B.p.	Bromide B.p.	Iodide B.p.	Thiouronium picrate (p. 35) M.p.	β-Naphthyl ether (p. 36) M.p.	β-Naphthyl ether picrate (p. 36) M.p.	Notes
Methyl	−24	4	43	224	72	118	
Ethyl	12	38	72	188	37	102	
Isopropyl	36	60	89	196	41	95	
Allyl	46	70	102	154	16	99	Unsaturated.
Propyl	46	71	102	177	40	81	
t-Butyl	51	72	98	151			
(±)-s-Butyl	67	90	120	166	34	86	
Isobutyl	68	91	120	167	33	84	
Butyl	77	101	130	177	33	67	
3-Methylbutyl (Isopentyl)	100	119	147	173	28	94	
Pentyl	107	128	155	154	25	66	
1-Chloro-2,3-epoxypropane (Epichlorohydrin)	116						See Table 21.

Table 22 (cont). Halides, alkyl mono-

	Chlo-ride B.p.	Bro-mide B.p.	Io-dide B.p.	Thio-uro-nium pic-rate (p. 35) M.p.	β-Naph-thyl ether (p. 36) M.p.	β-Naph-thyl ether picrate (p. 36) M.p.	Notes
Hexyl	134	156	180	157			
Cyclohexyl	142	166	180d		116		
Heptyl	160	178	203	142			
Benzyl	179	198	24*	187	102	122	* M.p.
Octyl	183	203	225	134			
2-Phenylethyl	190	218		139	70	83	
(±)-1-Phenylethyl	195	205		167			
o-Chlorobenzyl	214	102/9 mm		213			CrO$_3$ → o-chlorobenzoic acid, 142.
m-Chlorobenzyl	215	109/10 mm		200			CrO$_3$ → m-chlorobenzoic acid, 158.
o-Bromobenzyl	125/20 mm	31*	47*	222			* M.p. CrO$_3$ → o-bromo-benzoic acid, 150.
m-Bromobenzyl	119/18 mm	41*	42*	205			* M.p. CrO$_3$ → m-bromobenzoic acid, 155.
p-Chlorobenzyl	214	51*	64*	194			* M.p. CrO$_3$ → p-chloro-benzoic acid, 241.
	M.p.	M.p.	M.p.				
m-Nitrobenzyl	45	58	84				KMnO$_4$ → m-nitro-benzoic acid, 141.
o-Nitrobenzyl	48	46	75				KMnO$_4$ → o-nitro-benzoic acid, 146.
p-Bromobenzyl	50	63	80	219			CrO$_3$ → p-bromobenzoic acid, 251.
p-Nitrobenzyl	71	99	127				KMnO$_4$ → p-nitro-benzoic acid, 240.

Table 23. Halides, alkyl poly-

	B.p.	Thio-uronium picrate (p. 36)	β-Naph-thyl ether (p. 37)	Notes
Dichloromethane (Methylene dichloride)	42	267	133	
cis-1,2-Dichloroethylene	48			Dibromide, 192.
trans-1,2-Dichloroethylene	60			Dibromide, 192

89

Table 23 (cont). Halides, alkyl poly-

	B.p.	Thio-uronium picrate (p. 36)	β-Naphthyl ether (p. 37)	Notes
1,1-Dichloroethane	60		200	
(Ethylidene dichloride)				
Chloroform	61			Forms carbylamine with pri. amines and boiling ethanolic KOH.
Carbon tetrachloride	77			Forms carbylamine (as above).
1,2-Dichloroethane	83	260	217	
(Ethylene dichloride)				
Trichloroethylene	90			Chlorine atoms are unreactive.
Dibromomethane	97	267	133	
(Methylene dibromide)				
1,2-Dichloropropane	98	232	152	
(Propylene dichloride)				
1-Bromo-2-chloroethane	106		217	
1,1-Dibromoethane	112			
Tetrachloroethylene	121		117	Chlorine atoms are unreactive.
1,3-Dichloropropane	123	229	148	
(Trimethylene dichloride)				
1,2-Dibromoethane (m.p. 10)	132	260	217	
(Ethylene dibromide)				
1,2-Dibromopropane	141	232	152	
(Propylene dibromide)				
1,1,2,2-Tetrachloroethane	147			
(Acetylene tetrachloride)				
Bromoform	149			Forms carbylamine (see chloroform).
1,3-Dibromopropane	167	229	148	
(Trimethylene dibromide)				
Di-iodomethane	180		133	
(Methylene di-iodide)				
Benzylidene dichloride	212			Conc. H_2SO_4 at 50° → benzaldehyde, which may be characterized as 2,4-dinitrophenylhydrazone, 237.
(Benzal chloride)				
Trichloromethylbenzene	221			Boiling with aq. Na_2CO_3 → benzoic acid, 122.
(Benzotrichloride)				
1,5-Dibromopentane	221	247		
1,3-Di-iodopropane	224		148	
	M.p.			
1,2-Di-iodoethane	82	260	217	
(Ethylene di-iodide)				
Carbon tetrabromide	91			
Iodoform	119			Yellow. With quinoline in ether → compound, 65.
Hexachloroethane	187*			* Sublimes. Camphor-like odour.

Table 24. Halides, aryl

	B.p.	M.p.	Sulph-onamide (p. 37) position	Sulph-onamide (p. 37) M.p.	Nitro deriv. (p. 37) position	Nitro deriv. (p. 37) M.p.	meth-od	Notes
Fluorobenzene	85		4	125	4	27	ii	
o-Fluorotoluene	114		5	105				
m-Fluorotoluene	116		6	173				
p-Fluorotoluene	117		2	141				
Chlorobenzene	132		4	143	2,4	52	ii	
Bromobenzene	156		4	162	2,4	75	ii	
o-Chlorotoluene	159		5	126	3,5	64	ii	
m-Chlorotoluene	162		6	185	4,6	91	ii	
p-Chlorotoluene	162		2	143	2,6	76	i	
m-Dichlorobenzene	173		4	180	4,6	103	ii	
o-Dichlorobenzene	179		4	135	4,5	110	ii	
o-Bromotoluene	181		5	146	3,5	82	ii	
m-Bromotoluene	184		6	168	4,6	103	i	
p-Bromotoluene	185	28	2	165	2	47	iii	
Iodobenzene	188				4	174	i	Br_2 + Fe → p-bromo deriv., 91.
2,4-Dichlorotoluene	197		5	176	3,5	104	ii	
2,6-Dichlorotoluene	199		3	204	3,5	121	ii	
m-Iodotoluene	204				6	84	i	
o-Iodotoluene	211				6	103	ii	
p-Iodotoluene	211	35						Boiling HNO $\xrightarrow{3\,hr}$ acid, 265.
o-Chlorobenzyl chloride	214							Thiouronium picrate, 213 (p. 36); CrO_3 → o-chloro-benzoic acid, 142.
p-Chlorobenzyl chloride	214	29						Thiouronium picrate, 194; CrO_3 → p-chlorobenzoic acid, 242.
1,2,4-Trichlorobenzene	214	17	5	>200	3,5	103	ii	
1-Fluoronaphthalene	214							Picric acid deriv., 113 (p. 37).
m-Chlorobenzyl chloride	215							Thiouronium picrate, 200 (p. 36); CrO_3 → m-chlorobenzoic acid, 158.
m-Dibromobenzene	219		4	190	4,6	117	iii	
o-Dibromobenzene	224		4	175	4,5	114	ii	
1-Chloronaphthalene	260		4	186	4,5	180	iii	
1-Bromonaphthalene	281		4	192	4	85	iii	Picric acid deriv., 134.
1-Iodonaphthalene	305							Picric acid deriv., 128

Table 24 (cont). Halides, aryl

	B.p.	M.p.	Sulph-phonamide (p. 37) position	Sulph-phonamide (p. 37) M.p.	Nitro deriv. (p. 37) position	Nitro deriv. (p. 37) M.p.	meth-od	Notes
m-Bromobenzyl bromide		41						Thiouronium picrate, 205; $CrO_3 \rightarrow$ m-bromobenzoic acid 155.
p-Dichlorobenzene	174	53	2	180	2	56	iii	
1,2,3-Trichlorobenzene	218	53	4	227	4	56	ii	
2-Iodonaphthalene		55						Picric acid deriv., 95 (p. 37).
2-Bromonaphthalene		59	8	208				Picric acid deriv., 86 (p. 37).
2-Chloronaphthalene		60	8	232	1,8	175	ii*	* 6 hr. at 100°.
p-Bromobenzyl bromide		63						Thiouronium picrate, 219 (p. 36); $CrO_3 \rightarrow$ p-bromobenzoic acid, 251.
1,3,5-Trichlorobenzene		63	2	212d	2	68	ii	
p-Bromochlorobenzene		67			2	72	iii	
p-Dibromobenzene		87	2	195	2	84	iii	
1,3,5-Tribromobenzene		120	2	222d	2,4	192	ii	
1,2,3,4-Tetrachlorobenzene		139	3	99				
			3,6	227				

Table 25. Hydrocarbons

	B.p.	M.p.	Sulph-ona-mide (p. 38)	Nitro deriv. (p. 33) position	Nitro deriv. (p. 33) M.p.	meth-od	Picric acid deriv. (p. 27)	Notes
2-Methyl-1,3-butadiene (Isoprene)	34							Unsaturated; polymerizes easily; maleic anhydride adduct, 64 (p. 38).
Pent-1-yne	40							Unsaturated. Hg deriv., 118 (p. 38).
Cyclopentadiene	40							Unsaturated. Maleic anhydride adduct, 164 (p. 38). Forms dimer b.p. 170d, m.p. 32 on standing.
Penta-1,3-diene (Piperylene)	42							Unsaturated. Maleic anhydride adduct, 61 (p. 38).
Cyclopentene	44							Unsaturated.

Table 25 (cont). Hydrocarbons

	B.p.	M.p.	Sulph-ona-mide M.p. (p. 38)	Nitro deriv. (p. 33) posi-tion	M.p.	meth-od	Picric acid deriv. (p. 27)	Notes
Benzene	80	6	153	1,3	90	ii		
Cyclohexane	81	6						Oxidation with fuming HNO_3 → adipic acid, 153.
Cyclohexene	83							Unsaturated. Conc. HNO_3 → adipic acid, 153
Toluene	110		137	2,4	70	ii		
Ethylbenzene	136		109	2,4,6	37	ii		
p-Xylene	137	15	147	2,3,5	139	ii		
m-Xylene	139		137	2,4,6	182	ii		
Phenylacetylene	140							Unsaturated. Hg deriv., 125 (p. 38).
o-Xylene	144		144	4,5	71	ii		
Vinylbenzene (Styrene)	146							Unsaturated; polymerizes in the presence of a drop of H_2SO_4. Dibromide, 73.
Isopropylbenzene (Cumene)	153		105	2,4,6	109	ii		
α-Pinene	156							Unsaturated. Dibromide, 164.
Allylbenzene	157							Unsaturated. CrO_3 → benzoic acid, 122.
Propylbenzene	159		107					
1,3,5-Trimethylbenzene (Mesitylene)	165		142	2,4,6	235	i		
1,2,4-Trimethylbenzene (Pseudocumene)	168		181	3,5,6	185	ii		
Dicyclopentadiene	170d	32						Unsaturated. Benzo-quinone adduct, 157 (p. 38).
(+)-Limonene	176							Unsaturated; odour of lemons. Tetrabromide 104.
p-Isopropyltoluene (p-Cymene)	176		115	2,3,6	118	ii		
Dipentene ((±)-Limonene)	181							Unsaturated; odour of lemons. Tetrabromide, 124.
Indene	182							Unsaturated; polymerized by acid or heat. HNO_3 → phthalic acid, 195. Benzylidene deriv., 135 (p. 39).

93

Table 25 (cont). Hydrocarbons

	B.p.	M.p.	Sulph-ona-mide (p. 38)	Nitro deriv. (p. 33) posi-tion	Nitro deriv. (p. 33) M.p.	Nitro deriv. (p. 33) meth-od	Picric acid deriv. (p. 27)	Notes
Tetrahydronaphthalene (Tetralin)	207		135	5,7	95	1		
1-Methylnaphthalene	241			4	71	iii	141	
2-Methylnaphthalene	241	37		1	81	i	115	
Diphenylmethane	262	26		2,2′,4,4′	172	ii		CrO_3—H_2SO_4 → benzophenone, 48.
(−)-Camphene	160	51						Unsaturated; dibromide, 89.
Dibenzyl (Bibenzyl)	284	52		4,4′	180	i		CrO_3—H_2SO_4 → benzoic acid, 122.
Diphenyl (Biphenyl)	255	70		4,4′	234	iv		Br_2-acetic acid (boil for 2 hr) → 4,4′-dibromo deriv., 169.
1,2,4,5-Tetramethylbenzene (Durene)		79	155	3,6	205	ii	95	
Naphthalene		80		1	61	i	150	Odour of 'moth balls'; styphnic acid deriv., 168.
Acenaphthylene		92					201	Dibromide, 121.
Triphenylmethane		94		4,4′,4″	206	iii		
Acenaphthene		95		5	101	i	161	Styphnic acid deriv., 154.
Phenanthrene		100					144	CrO_3-acetic acid → quinone, 202. Styphnic acid deriv., 142.
2,3-Dimethylnaphthalene		104					124	Styphnic acid deriv., 149.
Fluoranthene		110					182	Styphnic acid deriv., 151.
2,6-Dimethylnaphthalene		111					143	Styphnic acid deriv., 159.
Fluorene		115		2	156	i	84*	Gives blue colour with conc. H_2SO_4; styphnic acid deriv., 134.
				2,7	199	ii		* Rather unstable.
trans-1,2-Diphenylethylene (trans-Stilbene)		124						Unsaturated; dibromide, 237, formed on warming with bromine. Styphnic acid deriv., 142 (p. 38).
Pyrene		150					227	Styphnic acid deriv., 191.
Anthracene		217					138	Maleic anhydride adduct, 263 (p. 38). CrO_3-acetic acid → quinone, 286. Styphnic acid deriv., 180 (p. 38).

Table 26. Ketones (C, H and O)

	B.p.	M.p.	2,4-Di-nitro-phenyl-hydra-zone (p. 39)	Semi-carba-zone (p. 39)	p-Nitro-phenyl-hydra-zone (p. 39)	Notes
Propanone (Acetone)	56		126	187	148	Monobenzylidene deriv., 42. Dibenzylidene deriv., 112 (p. 39).
Butan-2-one (Ethyl methyl ketone)	80		116	146	128	
But-3-en-2-one (Methyl vinyl ketone)	80			141		Unsaturated.
Butan-2,3-dione (Diacetyl)	88		315	235 mono 278 di	230 mono >310 di	Monobenzylidene deriv., 53 (p. 39).
2-Methylbutan-3-one (Isopropyl methyl ketone)	94		124 (118)	113	108	Benzylidene deriv., 117 (p. 39).
Pentan-3-one (Diethyl ketone)	102		156	139	144	Monobenzylidene deriv., 31.
Pentan-2-one (Methyl propyl ketone)	102		143	110	117	
3,3-Dimethylbutan-2-one (Pinacolone)	106		125	158	139	Benzylidene deriv., 41.
4-Methylpentan-2-one (Isobutyl methyl ketone)	117		95	130	79	
3-Methylpentan-2-one (s-Butyl methyl ketone)	118		71	94		
2,4-Dimethylpentan-3-one (Di-isopropyl ketone)	124		96 (107)	159*		* Varies with the rate of heating.
Hexan-2-one (Butyl methyl ketone)	128		106	122	88	
4-Methylpent-3-ene-2-one (Mesityl oxide)	130		203	164 (133)	133*	* Prepared without heating. Unsaturated.
Cyclopentanone	131		146	206	154	Benzylidene deriv., 190 (p. 39).
Methyl 2-oxopropionate (Methyl pyruvate)	136		187	208		See also *Table 19*.
2-Methylhexan-4-one (Ethyl isobutyl ketone)	136		75	152		
2-Methylhexan-3-one (Isopropyl propyl ketone)	136			119		
Pentan-2,4-dione (Acetylacetone)	139		209 di 122*	107*		* Pyrazole deriv. Oxime, 149 (p. 39). See also *Table 18*.
2-Methylcyclopentanone	139			184		
5-Methylhexan-2-one (Isopentyl methyl ketone)	144		95	143		

Table 26 (cont). Ketones (C, H and O)

	B.p.	M.p.	2,4-Dinitrophenylhydrazone (p. 39)	Semicarbazone (p. 39)	p-Nitrophenylhydrazone (p. 39)	Notes
Heptan-4-one (Dipropyl ketone)	144		75	133		
3-Hydroxybutan-2-one (Acetoin)	145		315	185		
Hydroxypropanone (Hydroxyacetone, Acetol)	146		129	196	173	
Heptan-2-one (Methyl pentyl ketone)	151		89	123	73	
2-Methylheptan-4-one (Isobutyl propyl ketone)	155			123		
Cyclohexanone	155		162	166	146	Benzylidene deriv., 118 (p. 39).
Ethyl 2-oxopropionate (Ethyl pyruvate)	155		155	206d		See also Table 19.
3,5-Dimethylheptan-4-one (Di-s-butyl ketone)	162			84		
2-Methylcyclohexanone	163		136	196	132	
4-Hydroxy-4-methylpentan-2-one (Diacetone alcohol)	165		203		209	
2-Oxopropionic acid (Pyruvic acid)	165d		218	222	220	See also Table 19.
2,6-Dimethylheptan-4-one (Di-isobutyl ketone)	168		92 (66)	122		
3-Methylcyclohexanone	168		155	191 (180)	119	Benzylidene deriv., 122 (p. 39).
4-Methylcyclohexanone	169		134	198	128	Benzylidene deriv., 99 (p. 39).
Methyl 3-oxobutanoate (Methyl acetoacetate)	170		119	152		Gives red colour with aq. $FeCl_3$. See also Tables 18 and 19.
2-Methylheptan-6-one (Isohexyl methyl ketone)	171		77	156		
Octan-2-one (Hexyl methyl ketone)	173		58	122	93	
Cyclohexyl methyl ketone	180		140	177	154	
Ethyl 3-oxobutanoate (Ethyl acetoacetate)	181d		96	133	218*	Gives red colour with aq. $FeCl_3$. See also Tables 18 and 19. * Pyrazole derivative.
Cycloheptanone	181		148	162	137	Benzylidene deriv., 108 (p. 39).
Nonan-5-one (Dibutyl ketone)	187		41	90		

Table 26 (cont). Ketones (C, H and O)

	B.p.	M.p.	2,4-Di-nitro-phenyl-hydra-zone (p. 39)	Semi-carba-zone (p. 39)	p-Nitro-phenyl-hydra-zone (p. 39)	Notes
Hexan-2,5-dione (Acetonylacetone)	190		256	185 mono 220 di	115	
Nonan-2-one	194		56	120		
Fenchone	194		140	183		Unsaturated.
Methyl 4-oxopentanoate (Methyl laevulinate)	196		141	144	136	
Cyclo-octanone	196		163	167		
2,6-Dimethylhept-2,5-dien-4-one (Phorone)	198	28	112	186		
α-Thujone	200		116	186		
β-Thujone	202		114	174		
Acetophenone	202	20	249	198	184	Benzylidene deriv., 58 (p. 39).
Ethyl 4-oxopentanoate (Ethyl laevulinate)	206		101	150	157	
(−)-Menthone	207		146	184		
Decan-2-one	209			124		
Decan-3-one	211			101		
Isophorone	214		130	191		
1-Phenylpropan-2-one (Benzyl methyl ketone)	216	27	156	190 (196)	145	
o-Methylacetophenone (Methyl o-tolyl ketone)	216		159	206		
Propiophenone	218	18	191	174	147	
o-Hydroxyacetophenone (o-Acetylphenol)	218	28	213	210		Oxime, 117 (p. 39). See also *Table 30.*
m-Methylacetophenone (Methyl m-tolyl ketone)	220		207	200		
Isopropyl phenyl ketone (Isobutyrophenone)	222		163	181		
Butyl phenyl ketone (Butyrophenone)	230		190	190		
p-Methylacetophenone (Methyl p-tolyl ketone)	223		260	205		
(+)-Pulegone	224		147	174		Unsaturated.
(+)-Carvone	225		190	162 (142)	174	Unsaturated.
1-Phenylbutan-2-one (Benzyl ethyl ketone)	226			135 (146)		
Undecan-2-one	228		63	122	90	

Table 26 (cont). Ketones (C, H and O)

	B.p.	M.p.	2,4-Di-nitro-phenyl-hydra-zone (p. 39)	Semi-carba-zone (p. 39)	p-Nitro-phenyl-hydra-zone (p. 39)	Notes
1-Phenylbutan-3-one	235		128	142		
m-Methoxyacetophenone	240			196		
Butyl phenyl ketone (Valerophenone)	242		166	166	162	
o-Methoxyacetophenone	245			183		Oxime, 83.
4-Oxopentanoic acid (Laevulinic acid)	245d	33	206	187d	175	Monobenzylidene deriv., 123 (p. 39). See also *Table 16.*
Ethyl acetonedicarboxylate	250		86	94		Red colour with aq. FeCl$_3$. See also *Table 19.*
α-Tetralone	257		257	226	231	Benzylidene deriv., 105 (p. 39).
p-Methoxyacetophenone	258	38	227	197	195	
Methyl 1-naphthyl ketone	298	34		235		Benzylidene deriv., 126 (p. 39).
Dibenzyl ketone	330	34	100	146		Monobenzylidene deriv., 162 (p. 39).
Benzylideneacetone	262	41	227	186	166	Unsaturated. Benzylidene deriv., 112 (p. 39).
Indanone		42	258	233	235	Benzylidene deriv., 113 (p. 39).
Diphenyl ketone (Benzophenone)		48	238	165	154	
Methyl 2-naphthyl ketone		53	262d	236		
Phenyl p-tolyl ketone		54 (60)	200	122		
Benzylideneacetophenone (Chalcone)		58	248 (208)	168 (180)		Unsaturated.
Benzyl phenyl ketone (Deoxybenzoin)		60	204	148	163	Benzylidene deriv., 102 (p. 39).
1-Phenylbutan-1,3-dione (Benzoylacetone)		61	151		101	Monobenzylidene deriv., 99. See also *Table 18.*
p-Methoxybenzophenone		62	180 228*		199	* From chloroform.
Fluorenone		83	284	234	269	Yellow.
Di-p-tolyl ketone		93	229	140		Oxime, 163 (p. 39).
Benzil		95	189	244d di 182 mono	290	Yellow.

Table 26 (cont). Ketones (C, H and O)

	B.p.	M.p.	2,4-Di-nitro-phenyl-hydra-zone (p. 39)	Semi-carba-zone (p. 39)	p-Nitro-phenyl-hydra-zone (p. 39)	Notes
m-Hydroxyacetophenone (m-Acetylphenol)		96	256	195		See also Table 30.
p-Hydroxyacetophenone (p-Acetylphenol)		110	261	199		See also Table 30.
Dibenzylideneacetone		112	180	190	173	Unsaturated.
p-Phenylacetophenone		121	241			Oxime, 187 (p. 39).
4-Hydroxy-3-methoxybenzyli-deneacetone (Vanillideneacetone)		130	230			Unsaturated.
Benzoylphenylmethanol (Benzoin)		133	245	206d		Acetyl deriv., 83.
p-Hydroxybenzophenone		135	242	194		
Furoin		136	216			Oxime, 161 (p. 39).
2,4-Dihydroxyacetophenone (Resacetophenone)		147	218	216		
p-Hydroxypropiophenone		148	229 240*			* From chloroform.
Furil		162	215		199d	
2,3,4-Trihydroxyacetophenone (Gallacetophenone)		173		225		Oxime, 163; triacetate, 85; see also Table 30.
(+)-Camphor		179	177	237	217	Benzylidene deriv., 98 (p. 39).

Table 27. Ketones (C, H, O and halogen or N)

	B.p.	M.p.	2,4-Di-nitro-phenyl-hydra-zone (p. 39)	Semi-carba-zone (p. 39)	p-Nitro-phenyl-hydra-zone (p. 39)	Notes
Chloroacetone	119		125	164*	83	* Variable.
αα-Dichloroacetone	120			163		
p-Fluoroacetophenone	196			219		Oxime, 80 (p. 39).
m-Chloroacetophenone	228			232	176	Oxime, 88 (p. 39).
p-Chloroacetophenone	232		236	201	239	
o-Aminoacetophenone	250d	20		290d		Oxime, 109 (p. 39); see also Table 9.

Table 27 (cont). Ketones (C, H, O and halogen or N)

	B.p.	M.p.	2,4-Di-nitro-phenyl-hydra-zone (p. 39)	Semi-carba-zone (p. 39)	p-Nitro-phenyl-hydra-zone (p. 39)	Notes
o-Nitroacetophenone	178/32 mm	27	154	210d		Oxime, 117 (p. 39).
p-Chloropropiophenone	134/31 mm	36	222	176		
Phenacyl bromide		50	220	146		
p-Bromoacetophenone	255	51	230	208	248	
Phenacyl chloride		59	212	156		
p-Chlorobenzophenone		78	185			Oxime, 163 (p. 39).
m-Nitroacetophenone		80	232	259		
m-Aminoacetophenone		99		196		See also Table 9.
p-Aminoacetophenone		106	259	250		See also Table 9.
p-Bromophenacyl bromide		108	218			Oxime, 115 (p. 39). Benzoic acid ester, 119.
p-Phenylphenacyl bromide		126	228			Benzoic acid ester, 167.
4,4'-Bis-(dimethylamino)benzo-phenone (Michler's ketone)		174	273			Oxime, 233 (p. 39). See also Table 12.

Table 28. Nitriles (Some nitriles are also listed in Table 16)

	B.p.	M.p.	Car-boxylic acid (p. 40)	p-Bromo-phenacyl ester of acid (p. 31)	Amide (p. 40)	Notes
Propenonitrile (Acrylonitrile)	78					Unsaturated. 2-Naphthol adduct, 142.
Acetonitrile	82			85		
Propionitrile	97			61		
2-Methylpropionitrile (Isobutyronitrile)	108			77		
Butanonitrile (Butyronitrile)	118			63		
But-3-enonitrile (Allyl cyanide)	118			60		Unsaturated.
Chloroacetonitrile	127				120	

Table 28 (*cont*). Nitriles

	B.p.	M.p.	Carboxylic acid (p. 40)	p-Bromophenacyl ester of acid (p. 31)	Amide (p. 40)	Notes
3-Methylbutanonitrile (Isovaleronitrile)	129			68		
Pentanonitrile (Valeronitrile)	140			75		
4-Methylpentanonitrile (Isocapronitrile)	155			77		
Hexanonitrile (Capronitrile)	163			72		
(±)-Mandelonitrile	170d	21	118	113		
Benzonitrile	191		122	119	128	Smell of bitter almonds.
o-Toluonitrile	205		104	57	142	
m-Toluonitrile	212		111	108	96	
p-Toluonitrile	218	29	180	153	160	
Malononitrile	219		133d		170	
Phenylacetonitrile (Benzyl cyanide)	232		76	89	157	
Adiponitrile	295		153	155	220	
1-Naphthonitrile	299	35	161	135	202	
m-Chlorobenzonitrile		41	158	117	134	
o-Chlorobenzonitrile		43	142	106	142	
Succinonitrile		54	185	211	260	
p-Chlorobenzonitrile		92	241	126	179	
Phthalonitrile		141	200d	153	220	
p-Nitrobenzonitrile		148	240	134	201	

Table 29. Nitro-, Halogenonitro-compounds and Nitro-ethers

	B.p.	M.p.	Nitro deriv. (p. 33) position	M.p.	method	Colour with aq. NaOH	Notes
Nitromethane	101						Acid to litmus; benzylidene deriv., 58 (p. 39).
Nitroethane	114						Benzylidene deriv., 64 (p. 39).
2-Nitropropane	120						Reduction with Sn + HCl → isopropylamine (p. 41).
1-Nitropropane	132						Immiscible with water. Sn + HCl → propylamine (p. 41).

Table 29 (cont). Nitro-, Halogenonitro-compounds and Nitro-ethers

	B.p.	M.p.	Nitro deriv. (p. 40) position	M.p.	meth-od	Colour with aq. NaOH	Notes
Nitrobenzene	211		1,3	90	ii		Pale yellow; odour of bitter almonds. Sn + HCl → aniline (p. 41).
o-Nitrotoluene	222		2,4	70	ii		Pale yellow; odour of bitter almonds. Sn + HCl → o-toluidine (p. 41).
1,3-Dimethyl-2-nitro-benzene (2-Nitro-m-xylene)	226	13	2,4,6	182	ii		
Phenylnitromethane	226d						Yellow; benzylidene deriv., 75.
m-Nitrotoluene	233	16					Pale yellow; Sn + HCl → m-toluidine (p. 41). Boiling aq. $K_2Cr_2O_7$—H_2SO_4 → acid, 140 (p. 23).
6-Chloro-2-nitrotoluene	238	37					Pale yellow. $K_2Cr_2O_7$—H_2SO_4 → acid, 161 (p. 23).
1,4-Dimethyl-2-nitro-benzene (2-Nitro-p-xylene)	240		2,3,5	139	ii		
p-Ethylnitrobenzene	241						Sn + HCl → p-ethylaniline (p. 41).
1,3-Dimethyl-4-nitro-benzene (4-Nitro-m-xylene)	244	2	2,4,6	182	ii		
o-Chloronitrobenzene	246	32	2,4	52	ii		Pale yellow.
1,2-Dimethyl-3-nitro-benzene (3-Nitro-o-xylene)	250	15	3,4	82	ii		Pale yellow.
1,2-Dimethyl-4-nitro-benzene (4-Nitro-o-xylene)	258	29	3,4	82	ii		
o-Methoxynitrobenzene (o-Nitroanisole)	265	9	2,4 / 2,4,6	88* / 68	i / ii		* Nitration at 0°.
o-Ethoxynitrobenzene (o-Nitrophenetole)	267		2,4 / 2,4,6	86* / 78	i / ii		* Nitration at 0°.
o-Nitrodiphenyl	320	37	2,4'	93	ii		
o-Bromonitrobenzene	259	41	2,4	72	ii		Pale yellow.
2-Nitro-1,3,5-trimethyl-benzene (Nitromesitylene)	255	44	2,4 / 2,4,6	86 / 235	iv / ii		

Table 29 (cont). Nitro-, Halogenonitro-compounds and Nitro-ethers

	B.p.	M.p.	Nitro deriv. (p. 40) position	M.p.	meth-od	Colour with aq. NaOH	Notes
m-Chloronitrobenzene	236	44	3,4	36	ii		Pale yellow; Sn + HCl → m-chloroaniline (p. 41).
p-Nitrotoluene	234	52	2,4	70	ii		Pale yellow; odour like nitrobenzene. $K_2Cr_2O_7$-dil. H_2SO_4 → acid, 241 (p. 23).
1-Chloro-2,4-dinitrobenzene		52	2,4,6	183	ii	Red → lilac	Reactive chlorine atom; boiling 2N NaOH → 2,4-dinitrophenol, 114. Hydrazine → 2,4-dinitro-phenylhydrazine, 199.
p-Methoxynitrobenzene (p-Nitroanisole)		53	2,4	87	i		Boiling conc. NaOH → p-nitrophenol, 114.
m-Bromonitrobenzene		56	3,4	59	ii		Pale yellow.
1,4-Dichloro-2-nitrobenzene		56	2,6	104	ii		Pale yellow. KOH in boiling aq. methanol → 4-chloro-2-nitromethoxy-benzene, 98.
β-Nitrovinylbenzene (β-Nitrostyrene)		58					Yellow. Sn + HCl → 2-phenylethylamine (p. 41).
p-Ethoxynitrobenzene		60	2,4 / 2,4,6	86 / 78	i / ii		Boiling with 40% HBr → p-nitrophenol, 114
1-Nitronaphthalene		60	1,3,8	218	ii		Yellow. Picrate, 71. CrO_3-acetic acid → 3-nitro-phthalic acid, 218.
2,6-Dinitrotoluene		66				Violet	Boiling dil. HNO_3 → acid, 202.
1-Methoxy-2,4,6-trinitro-benzene (2,4,6-Trinitroanisole)		68				Purple	Yellow. Boiling dil. NaOH → picric acid, 122. NH_3 in ethanol → picramide, 188.
2,4-Dinitrotoluene		70	2,4,6*	82	ii	Blue	CrO_3-conc. H_2SO_4 → acid, 183 (p. 23). * Explosive; not recommended.
1-Bromo-2,4-dinitrobenzene		72				Red	Pale yellow. Reactive bro-mine atom; boiling 2N NaOH → 2,4-dinitro-phenol, 114. Hydrazine → 2,4-dinitrophenyl-hydrazine, 199.
1,3-Dimethyl-5-nitro-benzene (5-Nitro-m-xylene)		75	4,5,6	125	ii		

Table 29 (cont). Halogenonitro-compounds and Nitro-ethers

	B.p.	M.p.	Nitro deriv. (p. 40)			Colour with aq. NaOH	Notes
			posi-tion	M.p.	meth-od		
1-Ethoxy-2,4,6-trinitro-benzene (2,4,6-Trinitrophenetole)		78				Red	Yellow. Boiling dil. NaOH → picric acid, 122.
2,4,6-Trinitrotoluene (T.N.T.)		82				Red	Explosive. Addition product with naphthalene, 97.
p-Chloronitrobenzene		83	2,4	52	ii		Pale yellow. Reactive chlorine atom; boiling aq. KOH → p-nitro-phenol, 114
1-Chloro-2,4,6-trinitro-benzene (Picryl chloride)		83				Red	Yellow. Reactive chlorine atom; warm aq. KOH → picric acid, 122. Naph-thalene adduct, 150.
1,4-Dibromo-2-nitro-benzene		84					Pale yellow. Reactive bro-mine atom; boiling aq. methanolic KOH → 4-bromo-2-nitromethoxy-benzene, 86.
1-Methoxy-2,4-dinitro-benzene (2,4-Dinitroanisole)		87	2,4,6	66	ii	Purple	Pale yellow. Boiling aq. KOH → 2,4-dinitro phenol, 114.
m-Dinitrobenzene		90				Purple	Pale yellow. Hot ethanolic NH_4SH → m-nitroani-line, 114 (p. 41).
1,3-Dimethyl-4,6-dinitro-benzene (4,6-Dinitro-m-xylene)		93	2,4,6	182	ii	Violet	Pale yellow. Hot ethanolic NH_4SH → 2,4-dimethyl-5-nitroaniline, 123 (p. 41).
p-Nitrodiphenyl		113	4,4'	233	ii		Carcinogenic.
o-Dinitrobenzene		118				None	Hot ethanolic NH_4SH → o-nitroaniline, 71; hot aq. NaOH → o-nitrophenol, 45.
1,3,5-Trinitrobenzene		122				Red	Pale yellow. Naphthalene adduct (in ethanol), 152
p-Bromonitrobenzene		126	2,4	72	ii		Pale yellow. Reactive bro-mine atom; boiling aq. KOH → p-nitrophenol, 114.
p-Dinitrobenzene		172				Green-yellow	Naphthalene adduct (in eth-anol), 118. Boiling ethan-olic NH_4SH → p-nitro-aniline, 147 (p. 41).

Table 30. Phenols (C, H and O)

Compound	B.p.	M.p.	FeCl₃ colour Aq.	FeCl₃ colour MeOH	Benzoate (p. 41)	Aryloxy-acetic acid (p. 42)	Toluene-p-sulphonate (p. 41)	3,5-Dinitro-benzoate (p. 42)	Notes
o-Cresol	190	31	B→G	G	Oil	152	55	138	p-Nitrobenzoate, 128 (p. 42). See also Table 4.
Salicylaldehyde	196		V	V		132	63		
m-Cresol	202	12	B→G	G	55	103	51	165	
p-Cresol	202	35	B	YG	71	136	70	188	
o-Methoxyphenol (Guaiacol)	205	30	R	G	57	116	82	142	
o-Ethylphenol	207		B	G	38	141		108	
2,4-Dimethylphenol (1,3-Xylen-4-ol)	211	27	B	GBt	38	142		164	
o-Acetylphenol (o-Hydroxyacetophenone)	218	28	VR	VR	87				See also Table 26.
m-Ethylphenol	216		V	G	52	75			p-Nitrobenzoate, 128 (p. 42). See also Table 19.
Methyl salicylate	224		V	V	92				p-Nitrobenzoate, 107 (p. 42). See also Table 19.
Ethyl salicylate	233		RV	V	79 (87)				
p-Isobutylphenol	236					125			
2-Methyl-5-isopropyl-phenol (Carvacrol)	237			Gt				80	
Isopropyl salicylate	238		V	V		151			Nitration (method ii, p. 34) → 3,5-dinitro deriv. 101. See also Table 19.
Propyl salicylate	239			Y					
m-Methoxyphenol	243		V	V		116			
p-Butylphenol	248	22	YG		27	81			p-Nitrobenzoate, 68 (p. 42).
4-Allyl-2-methoxyphenol (Eugenol)	253		YG	B	70	100*	85	131	* Hydrate, 81. Unsaturated.

Table 30 (cont). Phenols (C, H and O)

	B.p.	M.p.	FeCl₃ colour Aq.	FeCl₃ colour MeOH	Benzoate (p. 41)	Aryloxy-acetic acid (p. 42)	Toluene-p-sulphonate (p. 41)	3,5-Di-nitro-benzoate (p. 42)	Notes
Butyl salicylate	259 (270)		V	V					Nitration (method ii, p. 34) → 3,5-dinitro deriv., 61. See also *Table 19.*
Isobutyl salicylate	261		V	V					Nitration (method ii, p. 34) → 3,5-dinitro deriv., 72. See also *Table 19.*
2-Methoxy-4-(prop-1-enyl)-phenol (Isoeugenol)	267			Gt	106	94		158	Unsaturated. Dibromide, 94.
Phenol	182	42	V	G	69	99	96	146	p-Nitrobenzoate, 111 (p. 42). See also *Table 19.*
Phenyl salicylate (Salol)		42		VR	81				
p-Ethylphenol	219	47	B		60	97		132	
2,6-Dimethylphenol (1,3-Xylen-2-ol)	203	49	Y	Y	39	140		159	
3-Methyl-6-isopropyl-phenol (Thymol)	233	50	—	RBr	33	148	71	103	
p-Methoxyphenol	243	54	Vt	G	87	111			
o-Cyclohexylphenol		56			40				Nitration (method ii, p. 34) → 4,6-dinitro deriv., 106.
o-Phenylphenol		57 (67)			76	107	65		
3,5-Dihydroxytoluene, hydrate (5-Methylresorcinol, Orcinol)		58	BV	—	88	217		190	Heating at 100° → anhyd. form, 107.
3,4-Dimethylphenol (1,2-Xylen-4-ol)		62	B	Y	59	163		182	

106

Compound	M.P.							Notes
3,5-Dimethylphenol (1,3-Xylen-5-ol)	68	GB		24	111*	83	195	* Anhydrous; hydrate, 84.
2,4,5-Trimethylphenol (Pseudocumenol)	71	—		63	132			
2,5-Dimethylphenol (1,4-Xylen-2-ol)	75	YG		61	118		137	
2,3-Dimethylphenol (1,2-Xylen-3-ol)	75	B			187			p-Nitrobenzoate, 126 (p. 42).
4-Hydroxy-3-methoxy-benzaldehyde (Vanillin)	81	BV		78	188	115		See also Table 4.
p-t-Octylphenol	84	G		82				
p-Benzylphenol (p-Hydroxydiphenyl-methane)	84	G		87	108	75		
2-Hydroxybenzyl alcohol (Saligenin)	86			51	120			
p-t-Pentylphenol	92	Gt		61	192	55	217	* Colour of precipitate.
1-Naphthol (α-Naphthol)	94	Br	Pk*	56		89		
2-Naphthyl salicylate	95		V					Acetate, 136. See also Table 19. See also Table 26.
m-Acetylphenol (m-Hydroxyacetophenone)	96		V					
p-t-Butylphenol	100	G		82	86	110		
m-Hydroxybenzaldehyde	104	—	V	38	148			See also Table 4.
1,2-Dihydroxybenzene (Catechol)	105	G		84 di / 131 mono			152	p-Nitrobenzoate, 170 (p. 42).
3,5-Dihydroxytoluene, anhyd. (5-Methylresorcinol, Orcinol)	107	—	V	88	217		190	
p-Acetylphenol (p-Hydroxyacetophenone)	110	BrR	V	134				Acetate, 54. See also Table 26.

107

Table 30 (cont). Phenols (C, H and O)

	B.p.	M.p.	FeCl₃ colour Aq.	FeCl₃ colour MeOH	Benzo-ate (p. 41)	Aryloxy-acetic acid (p. 42)	Toluene-p-sulphonate (p. 41)	3,5-Di-nitro-benzo-ate (p. 42)	Notes
1,3-Dihydroxybenzene (Resorcinol)		110	V	G	117 di 135 mono	195	80	201	
Ethyl p-hydroxybenzoate		115	V	—	94				See also *Table 19.*
1,3,5-Trihydroxybenzene, dihydrate (Phloroglucinol)		117	V	G	174			162	Acetate, 104.
p-Hydroxybenzaldehyde		117	V	Y	91	198			See also *Table 4.*
2-Methylresorcinol (2,6-Dihydroxytoluene)		119	VBr	Br	106				
2-Naphthol (β-Naphthol)		123	W*	Gt	107	154	125	210	* Opalescent.
Methyl p-hydroxybenzoate		131	V	—	135				Acetate, 85. See also *Table 19.*
p-Cyclohexylphenol		132	R	G	118			168	p-Nitrobenzoate, 137 (p. 42).
1,2,3-Trihydroxybenzene (Pyrogallol)		133	R	G	90*	198		205	* Dibenzoate, 126; mono-benzoate, 138.
p-Hydroxybenzophenone		135	V*	Br	94				* Very weak. Acetate, 81.
2,4-Dihydroxybenzaldehyde (Resorcylaldehyde)		135	R	R					See *Table 4.*
1,2,4-Trihydroxybenzene (Hydroxyquinol)		140	R*		120				* In the presence of a trace of aq. NaOH. Acetate, 96.
2,4-Dihydroxyacetophenone		147							See *Table 26.*
p-Hydroxypropiophenone		148							See *Table 26.*
5-Methylsalicylic acid (p-Cresotic acid)		153	VB	B		185			Acetate, 152. See also *Table 16.*
3,4-Dihydroxybenzaldehyde (Protocatechualdehyde)		154			96				See also *Table 4.*

Compound	m.p.	Colour	Colour					Notes
3,5-Dihydroxybenzaldehyde	157							See Table 4.
Salicylic acid	158	V	V	132	191			Acetate, 135. See also Table 16.
2,3-Dihydroxynaphthalene	160	B*		152				Acetate, 105. * Precipitate also present.
3-Methylsalicylic acid (o-Cresotic acid)	163	VB	V		204			Acetate, 113. See also Table 16.
p-Phenylphenol	165	*	G	149	190	179		* Oxidized to p-benzoquinone.
1,4-Dihydroxybenzene (Quinol, hydroquinone)	171			200	250	159	317	
2,3,4-Trihydroxyacetophenone (Gallacetophenone)	173	Br	VBr					Triacetate, 85. See also Table 26.
1,4-Dihydroxynaphthalene	176			169				Acetate, 128.
4-Methylsalicylic acid (m-Cresotic acid)	177	V			165			Acetate, 139. See also Table 16.
2,7-Dihydroxynaphthalene	186			139	149	150		Acetate, 158. See also Table 16.
1-Hydroxy-2-naphthoic acid	195	B	BG					Acetate, 131. See also Table 16.
m-Hydroxybenzoic acid	200				206			
2,5-Dihydroxybenzoic acid (Gentisic acid)	200	BV	B					Diacetate, 118; 2-acetate, 172; 5-acetate, 131. See Table 16.
p-Hydroxybenzoic acid	215	O	O		278			Acetate, 187. See also Table 16.
1,3,5-Trihydroxybenzene, anhydrous (Phloroglucinol)	217	V	G	174* tri			162	Triacetate, 105. See also Table 18.
1,5-Dihydroxynaphthalene	250	—		235				* Di-, 126; mono-, 196. Acetate, 160.

Abbreviations for colours produced by ferric chloride: B, blue; Br, brown; G, green; O, orange; Pk, pink; R, red; t, transient; V, violet; W, white; Y, yellow; —, no colour.

Note. In the above ferric chloride tests, any deviation from the solvent stated will frequently invalidate the test.

Table 31. Phenols (C, H, O and halogen or N)

	B.p.	M.p.	FeCl₃ colour Aq.	FeCl₃ colour MeOH	Benzoate (p. 41)	Aryloxy-acetic acid (p. 42)	Toluene-p-sulphonate (p. 41)	3,5-Dinitro-benzoate (p. 42)	Notes
o-Chlorophenol	175	7	V	V	Oil	144	74	143	
o-Bromophenol	194	5	V	V	71	142	78		
3-Chloro-4-methylphenol	196			G		108		156	
m-Chlorophenol	214	33			71	109			
m-Bromophenol	236	33			86	108	53		
2-Bromo-4-chlorophenol	123/10 mm	33			99	139			
2,4-Dibromophenol	238	36	V	YG	98	153	120		
m-Iodophenol	217	40			73	115	60	183	
p-Chlorophenol		43	BV	G	86	156	71	186	
2,4-Dichlorophenol	209	43	VB	YG	97	140	125		
o-Iodophenol		43	—		34	135	80		
o-Nitrophenol	216	45	V	Br	59	156*	83	155	* Difficult to purify.
3-Methyl-4-nitrophenol		56			77				
4-Chloro-3-methylphenol		56 (66)	V	YG	86	178	98		
4-Chloro-2-isopropyl-5-methylphenol (p-Chlorothymol)		60	Y	Y				129	
p-Bromophenol		64	V	YG	102	157	94	191	Releases CO₂ from bicarbonate.
2,4,6-Trichlorophenol		68	—	—	76	182		136	Releases CO₂ from bicarbonate.
2,4,5-Trichlorophenol		68			93	157			
8-Hydroxyquinoline		75	BG	V	118	115	115		p-Nitrobenzoate, 174 (p. 42). See also Table 12.
p-Dimethylaminophenol		75	—			130	130		Acetate, 78. See also Table 12.
1-Bromo-2-naphthol		84	—	—	98				

110

Compound	M.p.	Colour with FeCl₃	Colour with FeCl₃					Remarks
m-Dimethylaminophenol	85			94		167		See also *Table 12*. Acetate, 95.
3,5-Dinitro-2-methylphenol	86			133		99		
p-Iodophenol	94		—	119	156	113	174	Releases CO_2 from bicarbonate.
2,4,6-Tribromophenol	95		—	81	200	113		
m-Nitrophenol	97	RV	V	95	155	97	159	See *Table 5*.
5-Bromosalicylaldehyde	106	V	Br	142				* Difficult to purify.
p-Nitrophenol	114	R	RBr	132	184	121	188	
2,4-Dinitrophenol	114	RBr	G	68	148*	103		
4-Chloro-3,5-dimethylphenol	115	BG		153	141			
m-Aminophenol	122	Br	RBr			110*	179	* N-Mono deriv, 157, O-mono deriv, 96. See *Table 9*.
2,4,6-Trinitrophenol (Picric acid)	122		—	163				Yellow; releases CO_2 from $NaHCO_3$; naphthalene adduct, 150.
p-Nitrosophenol	125d							See *Table 14*.
Salicylamide	139	R	V	143				O-Acetate, 138.
2-Amino-4,6-dinitrophenol (Picramic acid)	169	Br		220 N-		191		Red. See also *Table 9*.
o-Aminophenol	174	RBr	R	184		146* (139)		* N-Mono deriv.; O-mono deriv, 101.
2,4,6-Trinitroresorcinol (Styphnic acid)	179							Bright yellow. Naphthalene adduct, 168.
p-Aminophenol	184d	V	V → Br	234		168 di 252 N-	178	See also *Table 9*.
Pentachlorophenol	190		—	196	164	145		Acetate, 150. Releases CO_2 from bicarbonate.

Abbreviations for colours produced by ferric chloride: B, blue; Br, brown; G, green; O, orange; Pk, pink; R, red; t, transient; V, violet; W, white; Y, yellow; —, no colour.

Note. In the above ferric chloride tests, any deviation from the solvent stated will frequently invalidate the test.

Table 32. Quinones

	Colour	M.p.	Oxime (p. 42)	Semi-carbazone (p. 43) mono	di	Quinol (p. 43)	Notes
5-Isopropyl-2-methyl-1,4-benzoquinone (Thymoquinone)	Yellow	45	162	202d	237	143	
2-Methyl-1,4-benzoquinone (p-Toluoquinone)	Yellow	69	134d mono 220d di	178	240d	124	
2-Methyl-1,4-naphtho-quinone	Yellow	106	166 di 160 mono			170	
p-Benzoquinone	Deep yellow	115	144d mono 240d di	166	243	170	
1,4-Naphthoquinone (α-Naphthoquinone)	Yellow	125	198 mono 207d di	247		176	
1,2-Naphthoquinone (β-Naphthoquinone)	Red	146d	169 di 163 2- 109 1-	184		108*	* Anhydrous; hydrate, 60.
Quinhydrone	Dark green	171				170	$K_2Cr_2O_7$ + dil. $H_2SO_4 \rightarrow$ p-benzo-quinone.
9,10-Phenanthraquinone	Orange	206	158 mono 202d di	220d		148	
Acenaphthenequinone	Yellow	261	230 mono	192	271		
9,10-Anthraquinone	Pale yellow	286	224 mono			180	p-Nitrophenylhydra-zone, 238 (p. 24).

Table 33. Sulphonic acids and their derivatives

This table is arranged according to the boiling or melting point of the sulphonyl chloride because many of the acids do not have definite and reproducible values.

Sulphonyl chloride	M.p.	Acid	Amide (p. 43)	Ani-lide (p. 44)	Benzyl thiouro-nium salt of acid (p. 44)	Xan-thyl deriv. of amide (p. 44)	Notes
Methane-	*	†	90	99			* B.p. 161.
							† B.p. 167/10 mm.
Ethane-	*		58	58	115		* B.p. 177.
Propane-2-	*		60	84			* B.p. 61/9 mm.
Propane-1-	*	†	52				* B.p. 67/9 mm.
							† B.p. 136/1 mm.
o-Toluene-	10	57	156	136	170	183	
m-Toluene-	12		108	96			
Benzene-	14	66	155	110	148	206	Benzoyl deriv. of amide, 147.
2,5-Dimethylbenzene-	25	48*	148		184	176	* Dihydrate, 86.
2,4-Dimethylbenzene-	34	62	137	110	146	188	
2,5-Dichlorobenzene-	38	93	181	160	170		Acetyl deriv. of amide, 214.
3,4-Dimethylbenzene-	51	64	144		208	190	
p-Chlorobenzene-	53	68	144	104	175		
Toluene-2,4-di-	56		190	189			
2,4,6-Trimethylbenzene-	57	77	142	109		203	Acetyl deriv. of amide, 165.
Benzene-1,3-di-	63		229	144	214	170	
Naphthalene-1-	67	90	150	112	137		Benzoyl deriv. of amide, 194.
(+)-Camphor-10-	67	193	132	120	210		
p-Toluene-	69	92	137*	103	182	197	* Hydrate, 105.
p-Bromobenzene-	75	103	166	119	170		
Naphthalene-2-	76	91	217	132	191		Acetyl deriv. of amide, 145.
o-Carboxybenzene-	79	68 hyd. 134 anhyd.	222*	194	206		* Saccharin (sulpho-imide).
Naphthalene-2,7-di-	162		242		212		
Naphthalene-1,5-di-	183	245	310	249	253		
Anthraquinone-2-	197		261	193	211		
Anthraquinone-1-	214	218		216	191		
p-Hydroxybenzene- (Phenol-p-sulphonic)			177	141	169		Warm Br$_2$-water → tri-bromophenol, 95.
p-Aminobenzene- (Sulphanilic acid)		>300d	165*		182	208	Dibenzoyl deriv. of amide, 268.
							* See also *Table 10.*

113

Table 34. Thioethers (Sulphides)

	B.p.	M.p.	Sulphone (p. 45)	Notes
Dimethyl thioether	38		109	
Diethyl thioether	92		73	
Dipropyl thioether	142		29	
Di-isobutyl thioether	172		17*	* B.p. 265.
Dibutyl thioether	182		44	
Methylthiophenyl	188		88	
Ethylthiophenyl	204		41	
Diphenyl thioether	295		128	
Dibenzyl thioether	150	49	150	
Di-*p*-tolyl thioether	158	57	158	
Di-1-naphthyl thioether		110	187	
Di-2-naphthyl thioether		151	177	

Table 35. Thiols and Thiophenols

	B.p.	M.p.	2,4-Di-nitro-phenyl sulphide (p. 45)	H 3-nitro-phthal-oyl deriv. (p. 45)	3,5-Di-nitro-benzoyl deriv. (p. 45)	Notes
Methanethiol	6		128			
Ethanethiol	36		115	149	62	
Propane-2-thiol	58		95	145	84	
Propane-1-thiol	68		81	137	52	
Prop-2-en-1-thiol	90		72			
2-Methylpropane-1-thiol	88		76	136	64	
Butane-1-thiol	98		66	144	49	
3-Methylbutane-1-thiol	117		59	145	43	
Pentane-1-thiol	127		80	132	40	
Ethane-1,2-dithiol	146		248			
Hexane-1-thiol	111		74			
Cyclohexanethiol	159		148			
2-Hydroxyethanethiol	160		101			
Benzenethiol (Thiophenol)	169		121	130	149	
Propane-1,3-dithiol	169		194			
Heptane-1-thiol	176		82	132	53	
Toluene-ω-thiol	194		130	137	120	

Table 35 (cont). Thiols and Thiophenols

	B.p.	M.p.	2,4-Di-nitro-phenyl sulphide (p. 45)	H 3-nitro-phthal-oyl deriv. (p. 45)	3,5-Di-nitro-benzoyl deriv. (p. 45)	Notes
o-Toluenethiol (Thio-o-cresol)	194	15	101			
m-Toluenethiol (Thio-m-cresol)	195		91			
Octane-1-thiol	199		78			
Naphthalene-1-thiol (α-Thionaphthol)	209		176			
Decane-1-thiol	114/13 mm		85			
p-Toluenethiol	195	43	103			
p-Aminobenzenethiol (p-Aminothiophenol)		46				See *Table 10.*
p-Chlorobenzenethiol		53	123			
Naphthalene-2-thiol (β-Thionaphthol)		81	145			

115

INDEX

Acetals
  physical constants, 47
  preparation of derivatives, 22
  reactions, 9, 10
  solubility, 6
Acyl halides
  physical constants, 70–77
  reactions, 11
  solubility, 7
Alcohols
  physical constants, 47–50
  preparation of derivatives, 22
  reactions, 10
  solubility, 6
Aldehydes
  physical constants, 51–54
  preparation of derivatives, 23
  reactions, 9
  solubility, 6
Alkenes
  physical constants, 92–94
  reactions, 11
Alkyl halides
  physical constants, 88–90
  preparation of derivatives, 35
  reactions, 12
  solubility, 7
Alkynes
  physical constants, 92–94
  reactions, 11
Amides
  physical constants, 55, 56
  preparation of derivatives, 25
  reactions, 12, 13
  solubility, 6
Amides, N-substituted
  physical constants, 57
  preparation of derivatives, 26
  reactions, 13
  solubility, 6

Amines
  physical constants, 58–65
  preparation of derivatives, 26, 27
  reactions, 12
  solubility, 6
Amino-acids
  physical constants, 66, 67
  preparation of derivatives, 28
  reactions, 14
  solubility, 7
Ammonium salts
  reactions, 13
  solubility, 7
Anhydrides
  physical constants, 70–79
  reactions, 10, 11
  solubility, 6
Aryl halides
  physical constants, 91, 92
  preparation of derivatives, 37
  reactions, 12
  solubility, 7
Arylhydrazines
  physical constants, 68
  reactions, 12
  solubility, 7
Azo compounds
  physical constants, 68
  reactions, 13, 14
  solubility, 7
Azoxy compounds
  physical constants, 68
  reactions, 13, 14
  solubility, 7

Barfoed's reagent, 10
Boiling point determination, 5

Carbohydrates
  physical constants, 69

117